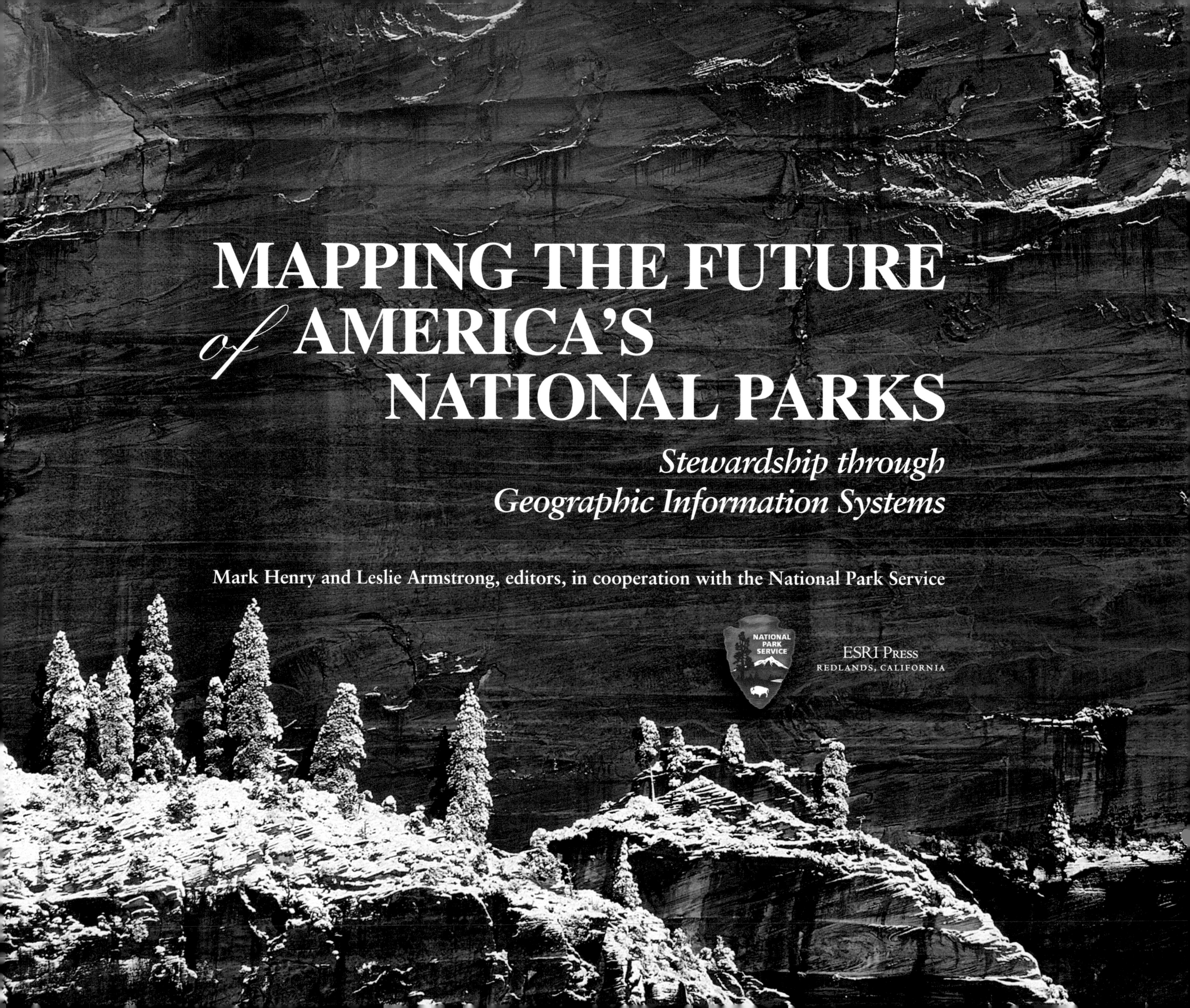

MAPPING THE FUTURE *of* AMERICA'S NATIONAL PARKS

Stewardship through Geographic Information Systems

Mark Henry and Leslie Armstrong, editors, in cooperation with the National Park Service

ESRI Press
REDLANDS, CALIFORNIA

First printing June 2004.

Printed in the United States of America.

Library of Congress Cataloging-in-Publication Data
Mapping the future of America's national parks : stewardship through geographic information systems / Mark Henry and Leslie Armstrong, editors, in cooperation with the National Park Service ; foreword by Allen Carroll.
p. cm.
ISBN 1-58948-080-5 (pbk. : alk. paper)
1. National parks and reserves—United States—Management. 2. Geographic information systems.
I. Henry, Mark, 1954– II. Armstrong, Leslie, 1956– III. United States. National Park Service.
SB486.M35M36 2004
917.3—dc22
2004008093

Published by ESRI, 380 New York Street, Redlands, California 92373-8100.

Books from ESRI Press are available to resellers worldwide through Independent Publishers Group (IPG). For information on volume discounts, or to place an order, call IPG at 1-800-888-4741 in the United States, or at 312-337-0747 outside the United States.

Contents

Foreword

National parks and the National Geographic Society were both born in the late nineteenth century in the waning years of a golden age of exploration. A new era of conservation dawned as the last bits of *terra incognita* disappeared from maps and people began to understand the limits of earth's natural bounties. Congress set aside Yellowstone and other early national parks as living remnants of the American wilderness and as sacred monuments to the vast wonders that fueled our country's growth and fired its imagination.

Gilbert Grosvenor, the legendary editor of *National Geographic,* played an instrumental role in creating the National Park Service. The magazine devoted its entire April 1916 issue to national parks, using photography, maps, and geographic analysis to set a precedent of public support and education. Four months later, President Woodrow Wilson signed the Organic Act establishing the National Park Service to oversee Yellowstone and thirty-six other national parks and monuments. As the twentieth century progressed, it became clear that simply setting aside parks wasn't enough. An increasingly complex interplay of natural systems and human activities required constant, active stewardship. The National Park Service faced an increasing array of difficult choices as the number of parks grew and a more affluent and mobile population visited in greater numbers. The service needed better and more scientific data and tools to address these challenges.

Today, the National Park Service makes management decisions and performs daily work using accurate and timely geographic information and advanced technology. Handsome hand-drawn maps that once documented discoveries during the Age of Exploration have exploded into geographic information systems in the Age of Information, transforming the way we manage national parks. Maps still document earth's resources and human activities, but they're now enriched with multiple levels of information. Software tools transform these information layers into data that reveal patterns, discern trends, and help us make better decisions. Maps explode with colorful images captured by orbiting spacecraft, blink with coordinates obtained by devices linked to global positioning satellites, and zoom instantaneously via fiber-optic cables to desktops around the world.

The National Park Service and National Geographic Society recently reaffirmed their nearly century-old commitment to preserve and explore our national parklands. As I write these words, I'm helping put the finishing touches on the eighth edition of the *National Geographic Atlas of the World.* I'm confident that the maps on its pages, like the stories, photographs, and maps in the book you are about to read, will inform and inspire. I'm hopeful, too, that both will open windows to our world's diversity, and perhaps even move us toward a commitment to preserve our planet's abundant natural and human treasures.

This book about GIS in our national parks shares the work of dedicated cartographers, GIS analysts, rangers, scientists, planners, and others engaged in a brand of stewardship not dreamed of in 1872, when Congress set aside Yellowstone "as a public park or pleasuring-ground for the benefit and enjoyment of the people." The maps on these pages amply demonstrate the power of geographic technology in our national parks. They show how maps vividly distill complex patterns to ease interpretation and help point the way toward informed decisions. But the final decisions, the important choices that will preserve or imperil our natural and cultural heritage, are human choices. And they're ultimately our choices as U.S. citizens. Here's hoping we have the insight to discern the subtle truths beneath these maps and the wisdom to protect our parks for future generations.

Allen Carroll
Chief Cartographer
National Geographic Society

Acknowledgments

The editors wish to thank the large cast of contributors, particularly the GIS specialists from the National Park Service and its partner agencies. Their abstracts on the NPS Web site at www.nps.gov/gis/mapbook served as the foundation for this book. The authors patiently answered questions and shared information about the evolving role and future of GIS in America's national parks. Their stories could not have been told without the invaluable assistance of the National Park Service book committee: Melia Lane-Kamahele, GIS regional coordinator for the Pacific Islands; Theresa Ely, GIS regional coordinator, Intermountain Region; Paul Voris, information technology specialist, Intermountain Region; Tennille Williams, cartographic student intern in the GIS Division; and David Duran, systems administrator in the GIS Division. Thanks also to the park and program managers who wrote the chapter introductions. Their expertise brought context and understanding to each chapter.

At ESRI Press, Michael Hyatt designed, produced, and copyedited the book and offered steady guidance. Savitri Brant designed the eye-catching cover. Jennifer Galloway edited many images. Thanks to Publisher Christian Harder for his skillful mentoring; Managing Editor Judy Hawkins for her suggestions; Michael Karman, David Boyles, Tiffany Wilkerson, and Edith M. Punt, for their encouragement and advice; Steve Hegle and Pam Spiva-Knutson, for their administrative support; and Cliff Crabbe, who oversaw print production. Special thanks to ESRI President Jack Dangermond, for his vision and leadership, and for creating an environment that made this book possible.

Introduction

". . . to conserve the scenery and the natural and historic objects and the wildlife therein and to provide for the enjoyment of the same in such manner and by such means as will leave them unimpaired for the enjoyment of future generations."

Organic Act, the legislated mission of the National Park Service, 1916

Americans identify with national parks by experiencing vistas and landscapes, by commemorating our heritage through preservation of sacred areas and historic battlefields. We are awed by the vastness of the Grand Canyon, or amazed by the graceful shapes of sandstone carved by the wind at Zion National Park. We take our children to Yellowstone National Park, perhaps to the very spot where our parents and their parents before them gazed upon Old Faithful, knowing it will be basically the same when our children arrive one day with families of their own, even after we are gone. This book tells an inspiring story about how we preserve these magnificent places through the use of technology and geographic data. Geographic information systems (GIS) and related technologies, such as global positioning systems (GPS), are the scientific basis and the necessary tools for upholding the mandate of the National Park Service to manage these American parks for our enjoyment, education, and research, and for future generations.

Often when we think about parks we think about exploring the great outdoors, enjoying the solitude and hoping to catch a glimpse of a bald eagle, a bighorn sheep, or an elephant seal. But many of our parks also tell the history of our country and preserve what we call cultural resources. Many American battlefields are designated as parks. We may feel solemn and reflective during a visit to Little Bighorn National Battlefield, where Lt. Col. George Armstrong Custer led five ill-fated companies of the U.S. Seventh Cavalry against the Sioux, Cheyenne, and Arapaho; or Antietam National Battlefield, where twenty-three thousand Americans were killed, wounded, or missing in a single day of the Civil War, many laid to rest in the adjacent national cemetery. At these and other national battlefields, we can trace the steps of opposing sides as they move over the hilly terrain and across fields of corn. We can imagine the volley of gunfire and cannonballs from strategic locations. We may recall the stories or service records of great-grandfathers and other ancestors who fought and died on these hallowed grounds. These human experiences relate to special events, places, and dates. They give us a sense of who we are, where we came from, and what our national ideals represent. We relate to the natural beauty and history that is preserved within park boundaries. Parks help us relate to ourselves, providing perspective on our life experiences.

These special places are unique from a geographic perspective, and that is often the fundamental reason for preserving them for all time. The National Park Service oversees a system of world-class wonders ranging over the entire geography of the United States, with 388 areas covering more than eighty-four million acres in forty-nine states, the District of Columbia, American Samoa, Guam, Puerto Rico, Saipan, and the Virgin Islands. The National

Park Service is a large land-management organization but is as uncommon as the parks it cares for. Parks are to be preserved in their most natural state and may even carry a wilderness designation that requires special care. How do we begin to understand the range of natural cycles in an ecosystem? How do we create or restore habitat for wolves or protect an endangered mushroom species from those who would gather it for large profit? When do we allow fire to take its natural course in a park? How do we build a road to improve visitor access and at the same time protect a nearby fragile wetland? How do we balance the preservation of prehistoric artifacts against the need for scientific study? Geographic analysis can provide or support solutions to all these questions. By examining patterns or trends in locations, activities, ground features, and associated time intervals, we can build campgrounds, identify and protect habitat, create fire management plans, and even suggest new commercial and military aircraft flight paths in ways to serve the mission of the National Park Service.

We can see that the use of geographic concepts is a natural step in the evolution of park management. This book seeks to extend our knowledge and experience with geographic technologies toward a model for global preservation for all who are interested in that goal. The analyses of incredible geographic data such as satellite and radar imagery, detailed terrain models, maps, and ground surveys help us better understand the national park system. From a low-flying aircraft, we can even reference video to positions on the earth and provide real-time information similar to what you see at an IMAX theater. We already use this technique to count birds or other animals and for inventories of tree and weed species.

Geography provides the reference framework, the lines of latitude and longitude, a unique position on the earth's surface from which we can observe and study parks, time and time again. We manipulate geographic information into models to help determine the historic landscape or predict the future alternatives. What did Yellowstone or Crater Lake look like before the cataclysmic eruptions that formed the calderas we see today? Can we predict what they will look like in the future? The modeling of landscapes, ecosystems, and phenomena such as natural wildfire cycles or prescribed-burn strategies helps us visualize different outcomes. Experts who study and manage wildfire attempt to do this every year, examining fuel models and fire history, revegetation plans, evacuation routes, and strategic protection of park resources and facilities. An examination of satellite imagery of Yellowstone just after the fires of 1989 and imagery from later dates reveals the rebirth of a rich and complex mosaic of vegetation. The young forests and riot of wildflowers instill in us the timelessness of nature.

Geographic information systems, global positioning systems, and their related technologies have already become important tools for educators, researchers, environmentalists, and other park advocates. GIS is a computer system with specialized software that portrays and compares complex human or natural activities and physical earth science data in a map or visual format. GIS allows us to analyze many layers of mapped data to discover how they interact or change one another. GPS, originally developed for military use, is a satellite-to-earth geometric measurement system used to calculate a precise location on the earth. These locations, or x,y measurements, represent features on the earth as points, lines, or polygons, and offer further analysis when we transfer them to a GIS. One can determine GPS locations of just about anything, from roads and bear den locations to the extent of a glacier or the outline of an archaeological structure. Some national parks, such as Mount Rainier and Glacier, use GPS equipment on snowplows to keep operators on the road and safe from falling down steep slopes.

Geographic data comes in many forms, such as three-dimensional data (x,y or horizontal measurement and z or vertical measurement), satellite imagery consisting of hundreds of bands of spectral information, and roads and trails data intelligently constructed in segments. Intelligent data construction enables the GIS operator to route and dispatch everyone from park maintenance workers to firefighters. Accurate geographic data is the basis and the most expensive part of any GIS.

Besides managing wildlife and other park resources, the National Park Service responds to the public with plans, information, and educational opportunities. The bureau is also responsible for the safety and enjoyment of millions of visitors every year. How does GIS assist with those duties? In the romantic "good old days," a president like Theodore Roosevelt would carve out a new expansive park from the Wild West based on politics and coarse land surveys. Today, in the global information era, the preservation of parks and the development of new parks require extensive study to justify federal expenditures and public benefit. The public demands participation in decisions about nearby parklands. GIS provides a way for us to analyze and visualize plans and alternatives displayed on maps, posters, and overhead computer projections. We use GIS to find lost hikers in our national parks and to revise trail maps and other features of interest. We don't have time to redraw maps with ruler and pen or the desire to lug armfuls of rolled maps and clear overlays to mark up changes presented at public meetings. The Internet provides geographic park information to the public for comment and examination of alternatives, costs and benefits of each, and virtual reality of what the area might look like for each alternative. Although geographic data and systems are employed for such studies, politics and other financial factors are often equally important and the use of technology is not always an acceptable method for those unfamiliar with its power.

Our goal is to let everyone use GIS rather than having to assign all GIS projects to park specialists. The National Park Service is developing a national GIS that brings together core data of all parks. We also see continued improvement of GIS in individual parks. Standard geographic databases and tools enable the public and park managers to access and use park information. This adds up to a more enjoyable and safer visitor experience too. Many park visitors already use GIS and GPS data for backcountry recreation, boating, fishing, driving tours, and park navigation. Perhaps virtual reality technology will soon let us present all park GIS data in a headset or other sensory device for a virtual park tour designed for those unable to travel to a park or for a congressional representative voting on the expansion of a park boundary or environmental regulations.

The National Park Service and partners involved with GIS and GPS present the following chapters in support of resource managers, law enforcement rangers, a wide range of other park staff, and the public. As the National Park Service develops and provides geographic information, the storehouse of legacy information supports scientific understanding about our changing park landscapes, ecosystems, and alternative management strategies. Through this work and use of geographic technologies, we strive to improve the success and health of parks around the globe.

Leslie Armstrong
GIS Program Coordinator
National Park Service

MAPPING THE FUTURE *of* AMERICA'S NATIONAL PARKS

Stewardship through Geographic Information Systems

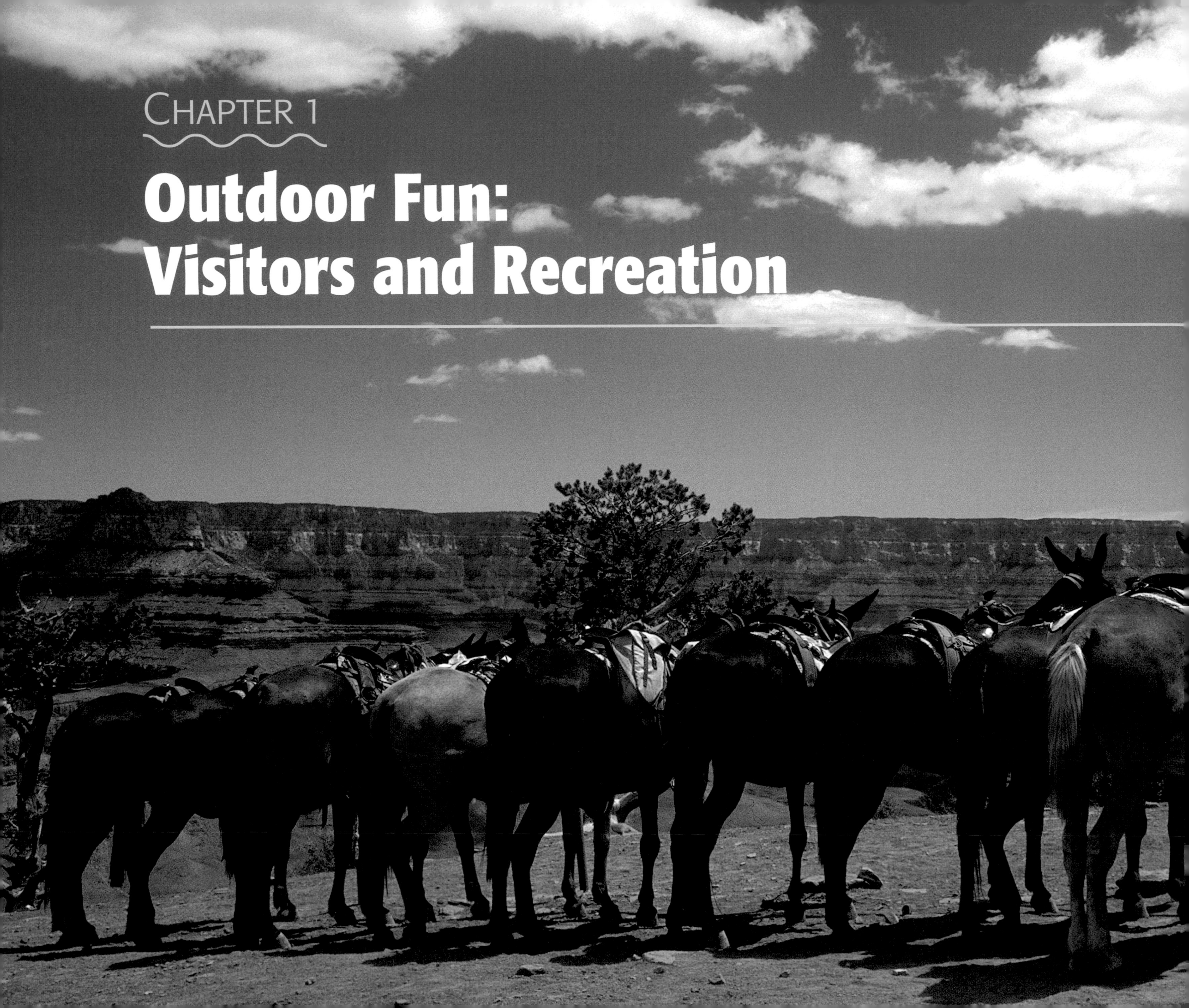

CHAPTER 1

Outdoor Fun: Visitors and Recreation

Who will gainsay that the parks contain the highest potentialities of national pride, national contentment, and national health? A visit inspires love of country; begets contentment; engenders pride of possession; contains the antidote for national restlessness. . . . He is a better citizen with a keener appreciation of the privilege of living here who has toured the national parks.

Stephen T. Mather, National Park Service director, 1917–1929

Park rangers and others who share their knowledge of our national parklands are always looking for ways to inspire visitors. This is a constant challenge for park interpreters, as visitors grow increasingly sophisticated in what they expect from gift shops, museums, trailside displays, brochures, and Web sites. Before they leave home, many visitors check the Internet for color images, maps, and other background on national parks. And when visitors arrive, they want clear yet visually exciting information to help them understand the trails, volcanoes, geysers, wildlife, cliff dwellings, climate, streams, campgrounds, and other resources they are about to explore.

Park interpreters now have a new and exciting tool to communicate a sense of place to park visitors. National parks, in partnership with the U.S. Geological Survey, have created GIS maps that offer stunning 3-D images of park topography and key features. At Grand Canyon National Park, for example, GIS maps help make sense of the immense scale and grandeur of geographic features that have a tendency to mystify and astound. Using satellite photography, GIS technology translates topography into layered maps that illustrate canyon views, rock layers, and how trails cross canyon walls. Yet GIS technology does more than make pretty maps. For example, GIS in the Grand Canyon shows the relationship between temperatures and canyon walls, information that could help save lives in a desert landscape with little or no water. At Hawaii Volcanoes National Park, maps use GIS layers and satellite imagery to create a bird's-eye view of the landscape. With this technology, serious volcano watchers and novices can instantly find what they need from the same map exhibit. The evolution of GIS has the potential to revolutionize Web sites, media, and other programs that help us interpret national parks. Technology now offers handheld computers for backcountry navigation, virtual hiking experiences, and GIS in park planning. The evolution of GIS will open up new exploration, such as virtual flying, as interpreters help visitors make the most of their vacations in our national parks.

Scenic Quality Analysis in the St. Croix River Valley

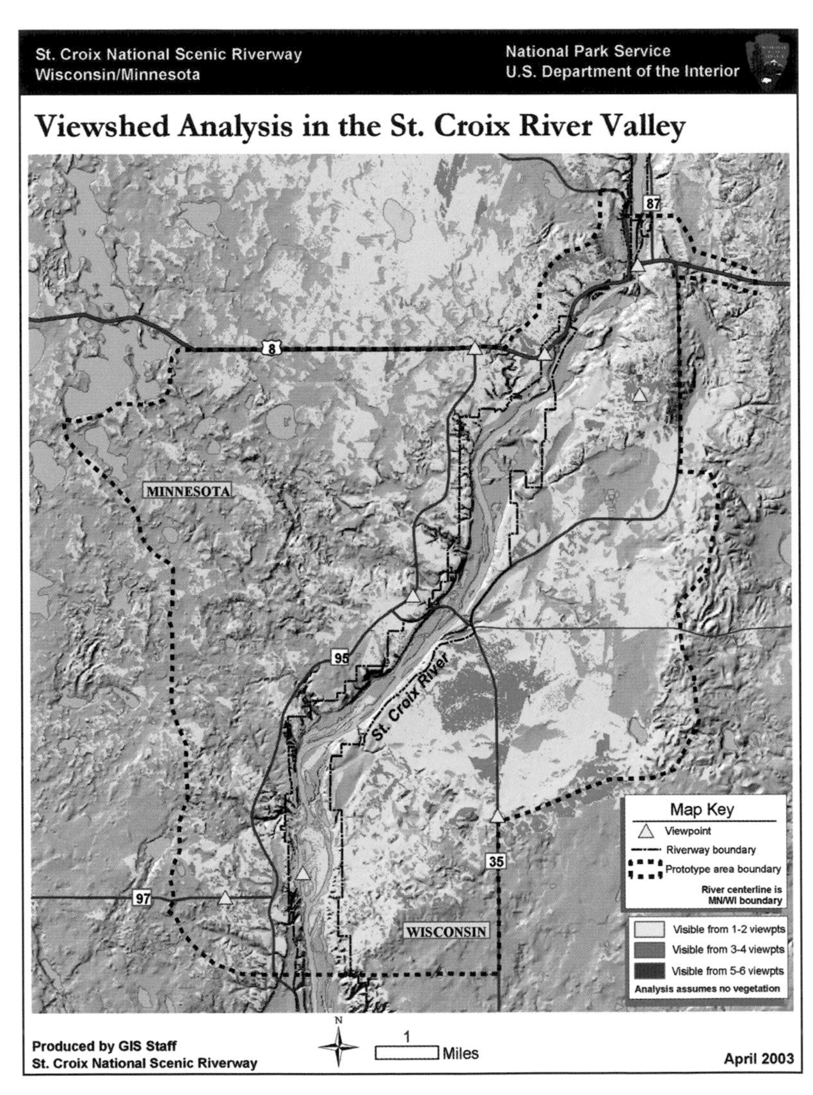

This GIS map displays areas visible to visitors from various scenic viewpoints along the St. Croix National Scenic Riverway in Minnesota and Wisconsin.

Gliding quietly in a canoe, you look up in awe at walls of ancient basalt towering over the river. Soon, the canyon opens into a wide valley carved by ancient glaciers. As you paddle, your gaze follows a great blue heron as it rises from the shoreline toward a bald eagle that dips and soars on the currents. This is the kind of view you would find during a visit to St. Croix National Scenic Riverway and its sister unit, the Lower St. Croix National Scenic Riverway, in Wisconsin and Minnesota. But idyllic scenes like this are facing the kind of development pressures that threaten many national parks today. These pressures take the form of new housing, gravel pits, commercial strips, and hobby farms and their accompanying roads, communication towers, power lines, and more. Development threatens to fragment habitat and destroy critical natural, cultural, and scenic resources beyond the boundaries of the St. Croix and Namekagon rivers as they flow to the Mississippi. To preserve the area, the nonprofit St. Croix Scenic Coalition, with the National Park Service acting in an advisory role, is using GIS software to identify and map valued community assets, such as scenic views. As a first step, area residents identified overlooks, trails, and roads that offered significant views in a particular area. The coalition downloaded the data into GIS software, which analyzed and displayed what could be seen from the different locations, and how often. Residents then returned to those locations and rated them using specific criteria. The resulting GIS map displayed the scenic quality of the area. These maps can help set priorities for conservation, direct development to other areas, and plan ecological corridors. The coalition hopes the effort will help reduce the harmful effects of rapid development. A community cannot protect scenic areas it has not identified. Preserving natural areas in the St. Croix River Valley serves the National Park Service mission to protect America's wonders for future generations.

Visitors can still admire this unimpeded view of the Lower St. Croix National Scenic Riverway.

St. Croix National Scenic Riverway
Wisconsin/Minnesota

National Park Service
U.S. Department of the Interior

Scenic Quality Analysis in the St. Croix River Valley

MINNESOTA
WISCONSIN
St. Croix River
87
8
95
97
35

See accompanying "ranking sheet" for information about evaluation criteria

Scenic Ranking
3 - 5
6 - 7
8 - 9
10 - 12
Higher values equate to higher scenic value

Map Key
Viewpoint
Riverway boundary
Prototype area boundary
River centerline is MN/WI boundary

Produced by GIS Staff
St. Croix National Scenic Riverway

N
1 Miles

April 2003

This ranking of scenic views shows the need to set priorities for preservation and protection from development pressures.

Hiking Trails of Great Falls, Maryland: Mapping for Visitor Safety

The view from the Potomac River toward Mather Gorge.

Visitors learn about the history of canal boats at Great Falls.

Hikers scramble over rocks and enjoy spectacular views in the Great Falls area of the Chesapeake & Ohio Canal National Historical Park within minutes of the nation's capital. The park attracts nearly a million visitors annually. Of the more than fifteen miles of hiking trails, a 1.7-mile-long section of the Billy Goat Trail features spectacular views of Mather Gorge below Great Falls. Yet the historic, scenic, and recreational features that draw hikers to one of the most difficult and heavily used trails on the middle Potomac River also have led to more injuries and missing hikers. To improve visitor safety and education, the park and its conservation partners upgraded trails and signs. They also spent two years creating a GIS-based map and safety education publication called "Hiking Trails of Great Falls Maryland." The two-sided brochure shows hiking trails on the front. The back offers visitor safety tips, trail descriptions, National Park Service regulations, emergency contact numbers, and alternative trails for less-strenuous hiking. The brochure also suggests ways for visitors to leave nature as they find it on trails. The map creators used various data collection and display software, including ESRI® ArcView® software, and global positioning systems. The park printed the brochure on water-resistant paper to meet the needs of visitors in most weather conditions and as a park keepsake.

Millions of visitors enjoy the natural, cultural, and recreational features at Chesapeake & Ohio Canal.

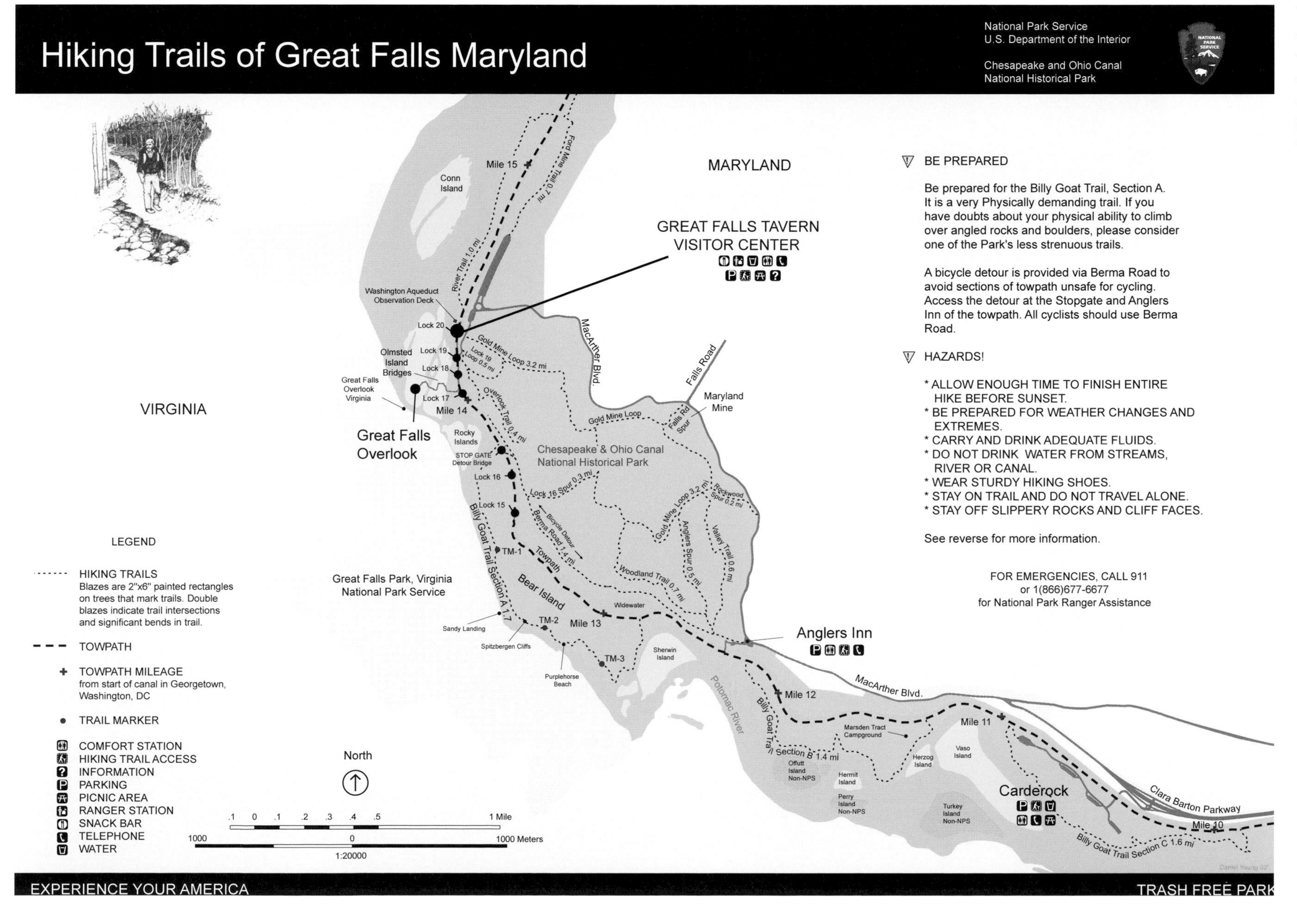
Hiking Trails of Great Falls Maryland
National Park Service
U.S. Department of the Interior
Chesapeake and Ohio Canal
National Historical Park
MARYLAND
VIRGINIA
GREAT FALLS TAVERN
VISITOR CENTER
Great Falls
Overlook
Chesapeake & Ohio Canal
National Historical Park
Great Falls Park, Virginia
National Park Service
Anglers Inn
Carderock
BE PREPARED
Be prepared for the Billy Goat Trail, Section A. It is a very Physically demanding trail. If you have doubts about your physical ability to climb over angled rocks and boulders, please consider one of the Park's less strenuous trails.
A bicycle detour is provided via Berma Road to avoid sections of towpath unsafe for cycling. Access the detour at the Stopgate and Anglers Inn of the towpath. All cyclists should use Berma Road.
HAZARDS!
* ALLOW ENOUGH TIME TO FINISH ENTIRE HIKE BEFORE SUNSET.
* BE PREPARED FOR WEATHER CHANGES AND EXTREMES.
* CARRY AND DRINK ADEQUATE FLUIDS.
* DO NOT DRINK WATER FROM STREAMS, RIVER OR CANAL.
* WEAR STURDY HIKING SHOES.
* STAY ON TRAIL AND DO NOT TRAVEL ALONE.
* STAY OFF SLIPPERY ROCKS AND CLIFF FACES.
See reverse for more information.
FOR EMERGENCIES, CALL 911
or 1(866)677-6677
for National Park Ranger Assistance
LEGEND
HIKING TRAILS
Blazes are 2"x6" painted rectangles on trees that mark trails. Double blazes indicate trail intersections and significant bends in trail.
TOWPATH
TOWPATH MILEAGE
from start of canal in Georgetown, Washington, DC
TRAIL MARKER
COMFORT STATION
HIKING TRAIL ACCESS
INFORMATION
PARKING
PICNIC AREA
RANGER STATION
SNACK BAR
TELEPHONE
WATER
North
1 Mile
1000 Meters
1:20000
Conn Island
Mile 15
Mile 14
Mile 13
Mile 12
Mile 11
Mile 10
Washington Aqueduct Observation Deck
Lock 20
Lock 19
Lock 18
Lock 17
Lock 16
Lock 15
Olmsted Island Bridges
Great Falls Overlook Virginia
Rocky Islands
STOP GATE Detour Bridge
Bear Island
Widewater
Sandy Landing
Spitzbergen Cliffs
Purplehorse Beach
Sherwin Island
Potomac River
Maryland Mine
Falls Road
MacArthur Blvd.
Clara Barton Parkway
Marsden Tract Campground
Herzog Island
Vaso Island
Offutt Island Non-NPS
Hermit Island
Perry Island Non-NPS
Turkey Island Non-NPS
EXPERIENCE YOUR AMERICA
TRASH FREE PARK

Visitor safety is a key goal of this brochure. It displays trails and recreational opportunities, offers tips for hiking preparedness, and lists precautions to take, especially along the Billy Goat Trail.

Personal Watercraft Environmental Assessment: Resource Mapping and Analysis

Whales, nesting ospreys, turtles, dolphins, and sea grass are all vulnerable to personal watercraft that skim the surface waters of our national parks. Noise and vessel maneuvers easily frighten shorebirds. In shallow water, an accelerating watercraft can blow out patches of sea grass that support marine life and estuaries. And it is hard to see, much less avoid, a turtle or other sea creature from a personal watercraft going 45 miles an hour. To study the effects of personal watercraft, Gulf Islands National Seashore used GIS software to map areas within view or earshot of personal watercraft areas. The park in Florida and Mississippi wanted the maps for public education, to understand how natural resources overlapped, and to identify areas where personal watercraft did not threaten the ecosystem. The maps displayed nesting sites of ospreys, eagles, herons, and shorebirds. The maps also showed popular areas for personal watercraft, swimming, and fishing, and displayed their locations in relation to documented sightings of air-breathing sea animals, such as endangered sea turtles, manatees, whales, and dolphins. The maps continue to help the park as it balances wildlife and habitat protection with recreational opportunities for park visitors.

These four maps show where the use of personal watercraft could affect bird, fish, and turtle nesting areas, as well as sea-grass habitat, in various parts of Gulf Islands National Seashore.

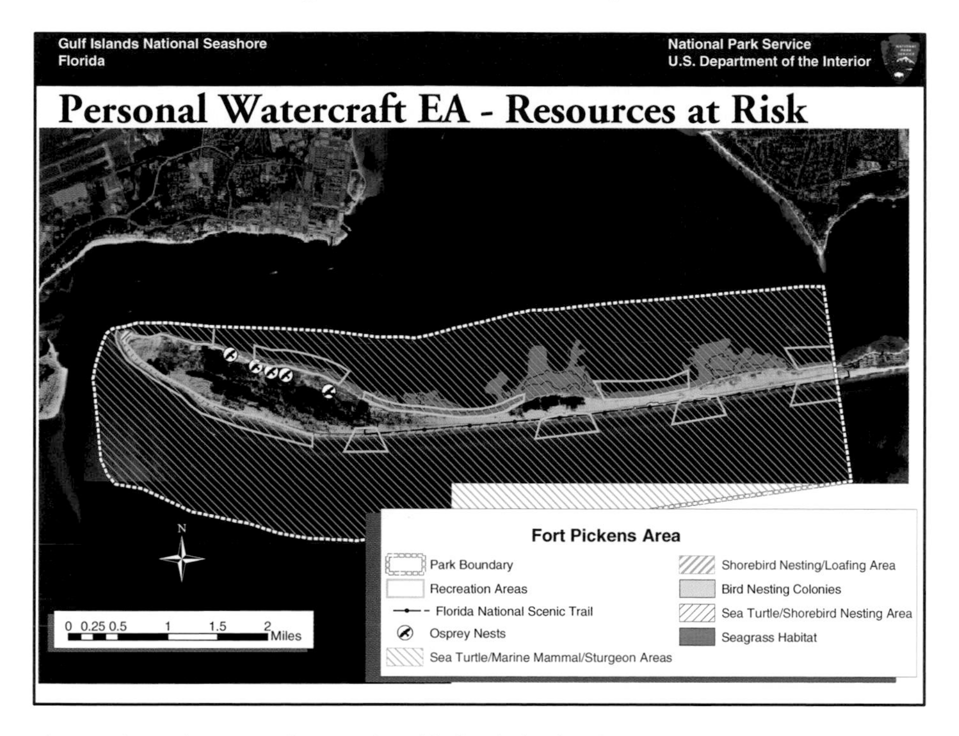

This map shows where personal watercraft could affect bird and turtle nesting areas and sea-grass habitat in the Fort Pickens area, Florida.

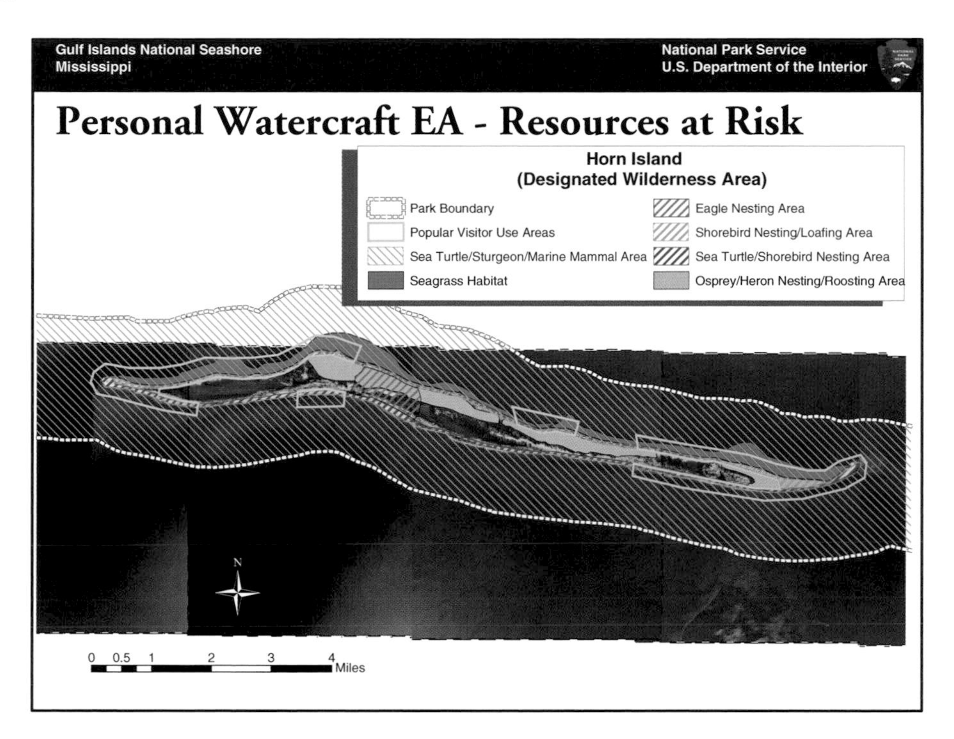

Resource map of Horn Island, Mississippi.

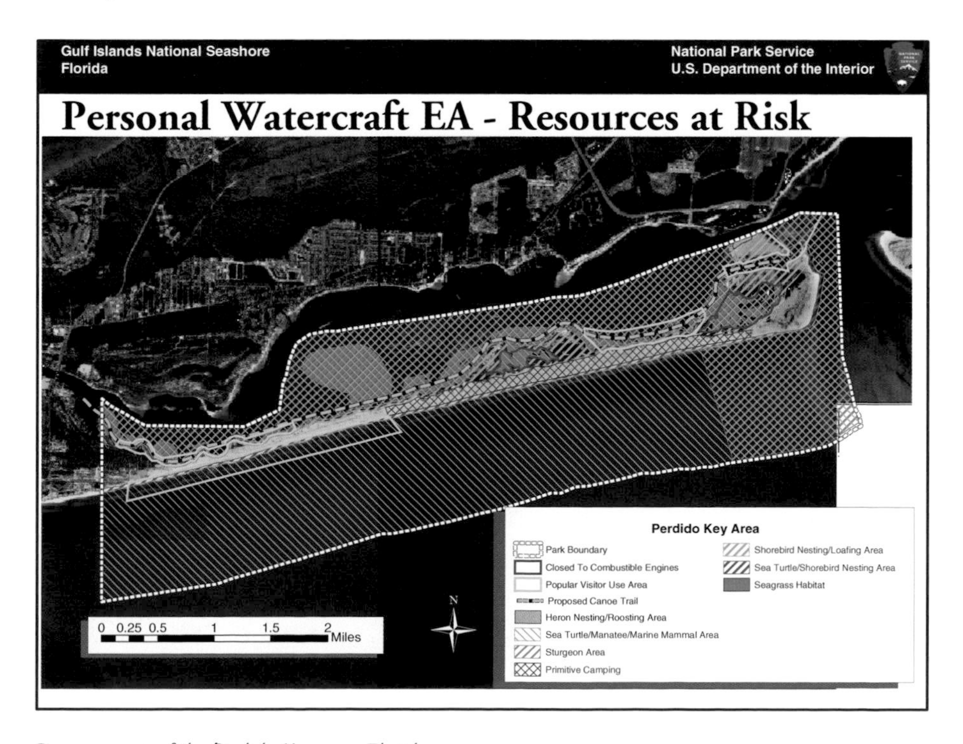

Resource map of the Perdido Key area, Florida.

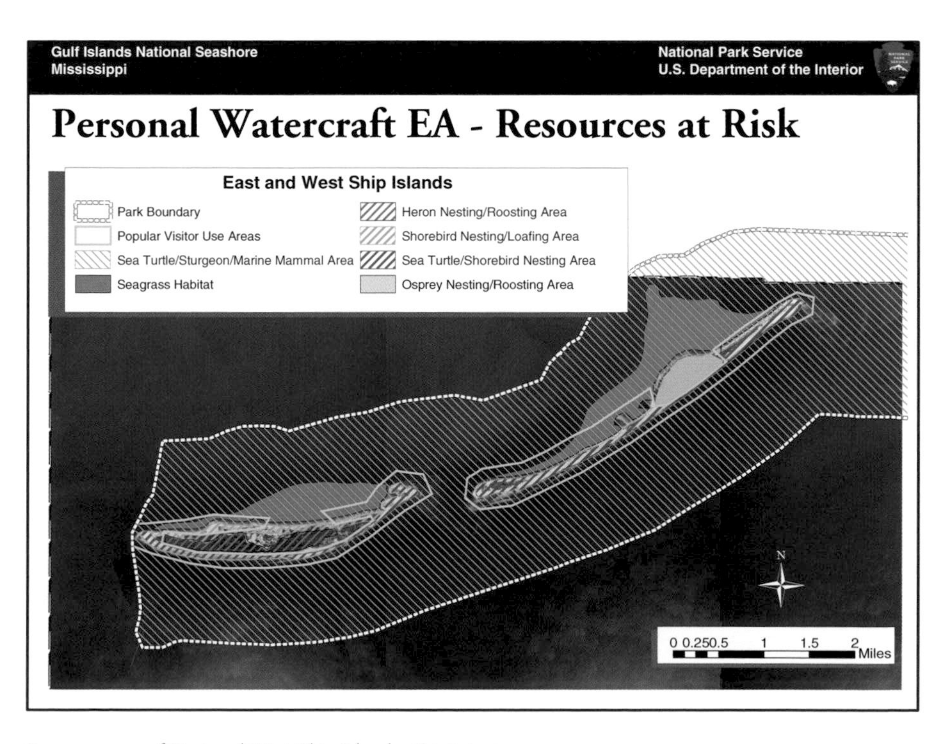

Resource map of East and West Ship Islands, Mississippi.

Use of GIS for Visitor Information at Dry Tortugas National Park

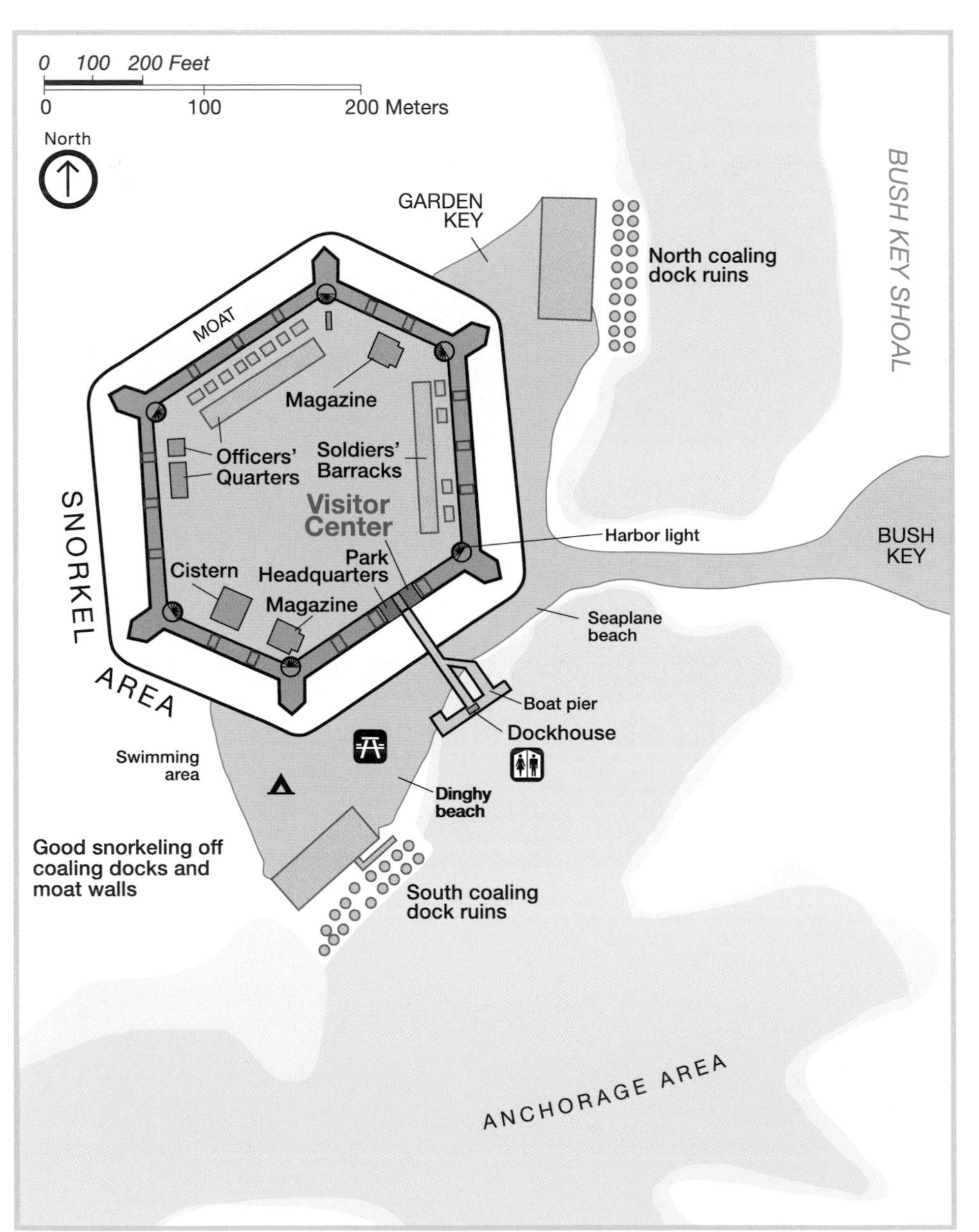

The seven islands that make up Dry Tortugas National Park offer stunning underwater views of shipwrecks, fish, coral, and other sea life. President Franklin D. Roosevelt set aside the pristine sands, shoals, and reefs as a national monument in 1935. The area seventy miles west of Key West, Florida, has enjoyed national park status since 1992. The park service has surveyed more than fifty square miles of seabed for display on GIS maps to benefit park managers, researchers, and the public. GIS mapping helps park managers identify and select scenic diving and snorkeling areas for public enjoyment and long-term study. One underwater area known as the Windjammer Site offers stunning views of a shipwrecked iron-hulled sailing ship that sank in 1906. Visitors can find a GIS map of the site at the park, on the Internet, and in a laminated version for underwater use during visits. The underwater trail guide and fish-watchers guides enhance the underwater experience. The park envisions adding historical and archaeological information with photographic and video imagery to make visits even more special. The park also is experimenting with ways to bring the experience to nondivers, children, and physically challenged people who want to know more about historic shipwrecks and the diverse natural environment around them.

Divers enjoy stunning underwater views at the Windjammer Site.

When work commenced on Fort Jefferson in 1846, it was to be the largest U.S. coastal fort to that time. The invention of the rifled cannon made the fort obsolete and it was never finished. President Roosevelt set aside the fort and surrounding waters as a national monument in 1935. The area was redesignated a national park in 1992.

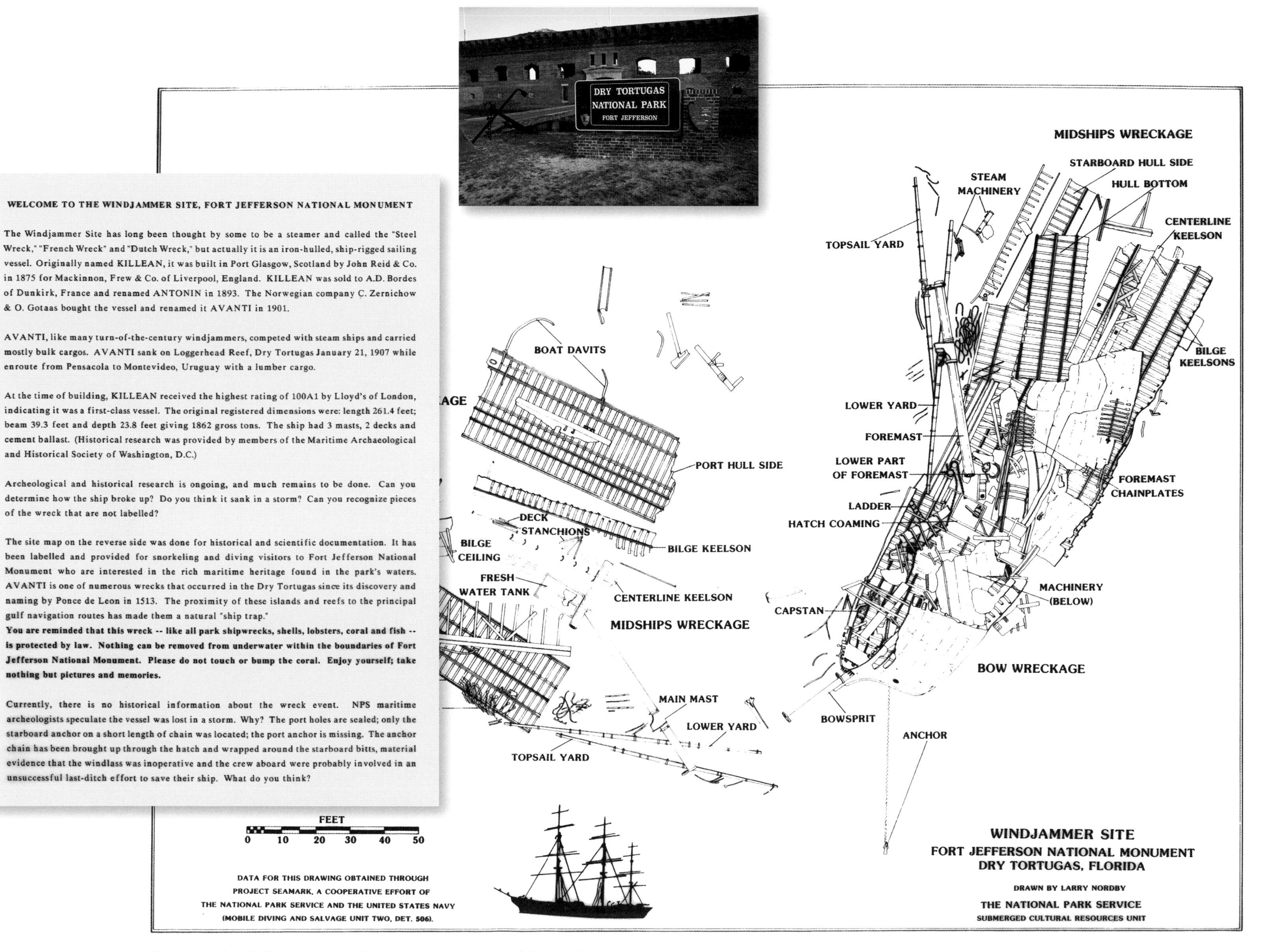

WELCOME TO THE WINDJAMMER SITE, FORT JEFFERSON NATIONAL MONUMENT

The Windjammer Site has long been thought by some to be a steamer and called the "Steel Wreck," "French Wreck" and "Dutch Wreck," but actually it is an iron-hulled, ship-rigged sailing vessel. Originally named KILLEAN, it was built in Port Glasgow, Scotland by John Reid & Co. in 1875 for Mackinnon, Frew & Co. of Liverpool, England. KILLEAN was sold to A.D. Bordes of Dunkirk, France and renamed ANTONIN in 1893. The Norwegian company Ç. Zernichow & O. Gotaas bought the vessel and renamed it AVANTI in 1901.

AVANTI, like many turn-of-the-century windjammers, competed with steam ships and carried mostly bulk cargos. AVANTI sank on Loggerhead Reef, Dry Tortugas January 21, 1907 while enroute from Pensacola to Montevideo, Uruguay with a lumber cargo.

At the time of building, KILLEAN received the highest rating of 100A1 by Lloyd's of London, indicating it was a first-class vessel. The original registered dimensions were: length 261.4 feet; beam 39.3 feet and depth 23.8 feet giving 1862 gross tons. The ship had 3 masts, 2 decks and cement ballast. (Historical research was provided by members of the Maritime Archaeological and Historical Society of Washington, D.C.)

Archeological and historical research is ongoing, and much remains to be done. Can you determine how the ship broke up? Do you think it sank in a storm? Can you recognize pieces of the wreck that are not labelled?

The site map on the reverse side was done for historical and scientific documentation. It has been labelled and provided for snorkeling and diving visitors to Fort Jefferson National Monument who are interested in the rich maritime heritage found in the park's waters. AVANTI is one of numerous wrecks that occurred in the Dry Tortugas since its discovery and naming by Ponce de Leon in 1513. The proximity of these islands and reefs to the principal gulf navigation routes has made them a natural "ship trap."
You are reminded that this wreck -- like all park shipwrecks, shells, lobsters, coral and fish -- is protected by law. Nothing can be removed from underwater within the boundaries of Fort Jefferson National Monument. Please do not touch or bump the coral. Enjoy yourself; take nothing but pictures and memories.

Currently, there is no historical information about the wreck event. NPS maritime archeologists speculate the vessel was lost in a storm. Why? The port holes are sealed; only the starboard anchor on a short length of chain was located; the port anchor is missing. The anchor chain has been brought up through the hatch and wrapped around the starboard bitts, material evidence that the windlass was inoperative and the crew aboard were probably involved in an unsuccessful last-ditch effort to save their ship. What do you think?

This map displays the location of pieces of the Windjammer, an iron-hulled sailing ship that sank in 1907.

CHAPTER 2

Into the Wild: Park Wilderness

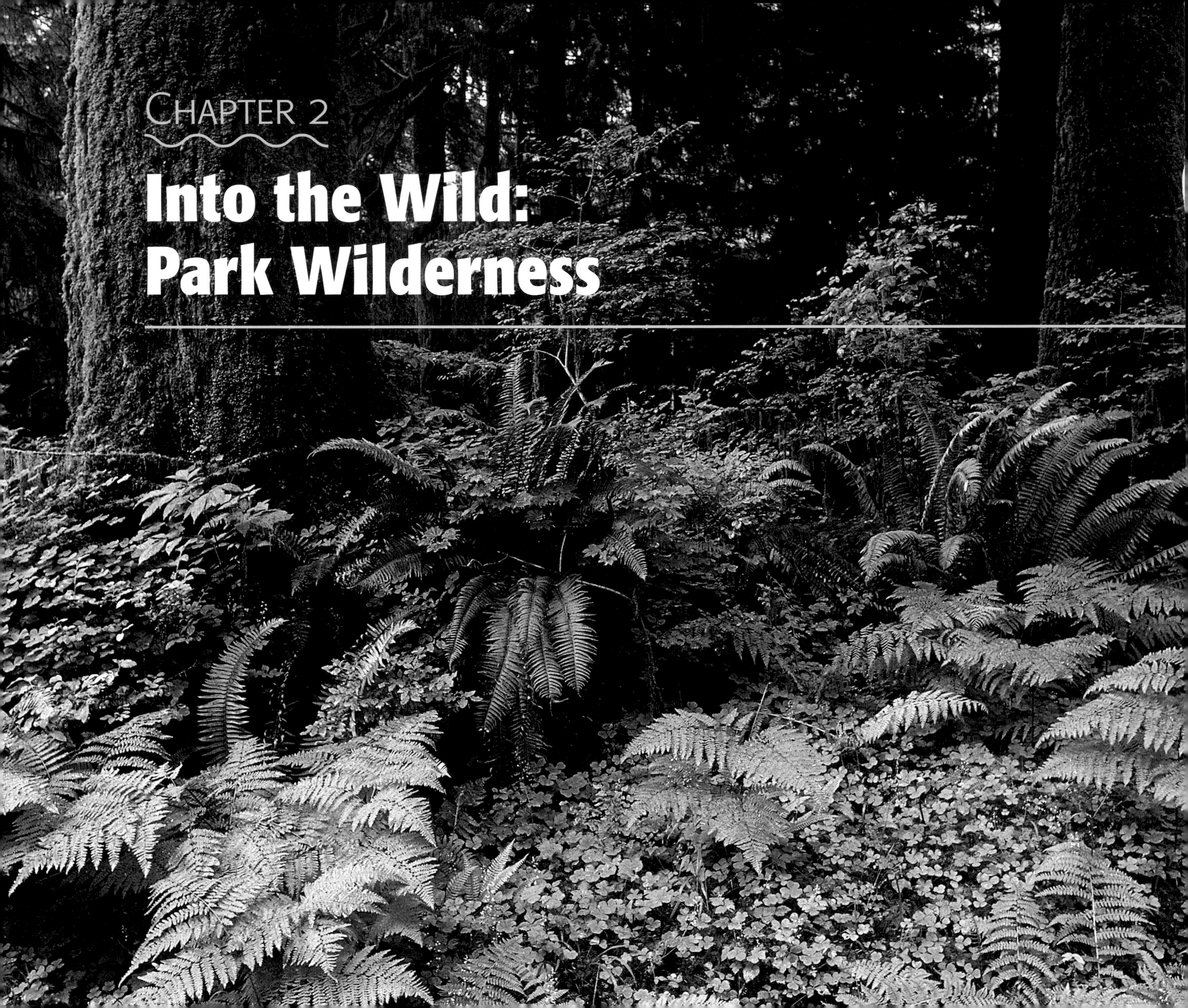

Without enough wilderness America will change. Democracy, with its myriad personalities and increasing sophistication, must be fibred and vitalized by the regular contact with outdoor growths—animals, trees, sun warmth, and free skies—or it will dwindle and pale.

Walt Whitman

The federal Wilderness Act of 1964 created an entirely new kind of place on the American landscape. The law set aside "wilderness" for preservation and protection from all kinds of development. Today, federal agencies manage nearly 106 million acres in the National Wilderness Preservation System. Of that total, the National Park Service manages more than 44 million acres. It also manages an additional 25 million acres of park service lands either proposed for wilderness or suitable for scientific study.

The wildest lands of our national parks pose a challenge for managers. The law severely limits what the park service and public can do in wilderness, compared to other public lands. The law prohibits driving and most motorized equipment in wilderness. But finding ways to preserve the character of these fragile lands while also leaving them untrammeled can frustrate park managers. For example, a wilderness designation can prevent managers from using chain saws to ease the fire threat in an overgrown forest or stop them from ordering bulldozers to clear firebreaks. In this environment, GIS and its suite of tools offer innovative ways to protect the "primeval character and influence" of our nation's wild lands. The technology improves the quality and speed of information about the landscape without harming it.

Wilderness managers, including rangers, natural and cultural resources specialists, and fire managers, need to know the location and condition of plant and animal populations, historic and prehistoric sites, campsites, trails and bridges, even trash piles. Scientists increasingly use wilderness areas to conduct research that requires pristine lands. GPS in the field, satellite remote sensing, and GIS on the computer allow most of this monitoring to take place without scarring wild lands. Knowing the precise location of wilderness boundaries, particularly when drawn close to park developments, is crucial for park planners. And GIS can target the location of fire management activities to lessen their impact on wilderness.

It is an irony, and one by no means appreciated universally, that many wilderness areas will need high levels of knowledge and understanding if they are to survive reasonably intact within the matrix of continued changes to our landscape. The National Park Service will call upon sophisticated tools, GIS among them, to protect these precious areas for our future enjoyment, benefit, study, and understanding.

Backcountry Management Planning in Alaska's National Parks

The National Park Service will produce more than eighty GIS maps to guide its management of Alaska's backcountry in Denali National Park and Preserve, Gates of the Arctic National Park and Preserve, Glacier Bay National Park and Preserve, and Wrangell-St. Elias National Park and Preserve. By definition, the backcountry in these northern parklands is undeveloped, with no major facilities or roads. Over the next several years, management plans created for these parks will address how to protect these exceptionally wild lands. They will also decide how and where to provide public access and use of these nationally significant areas. Together, these parks represent 26 percent of the country's national parklands and nearly 20 percent of all lands designated as wilderness. The park service plans to coordinate the development of the plans to encourage public involvement and allow consistent interpretation of federal laws and policies for the backcountry. The coordinated approach should result in options that reflect visitor use and backcountry protection from a regional perspective. The park service already has presented several GIS maps, including one of Gates of the Arctic that shows the park in relation to neighboring lands and the types of recreation in the region. GIS technology will be a critical tool in evaluating plan alternatives and environmental impacts in the backcountry.

The wilds of Alaska draw visitors to Glacier Bay National Park and Preserve, at left, and Gates of the Arctic National Park and Preserve, above.

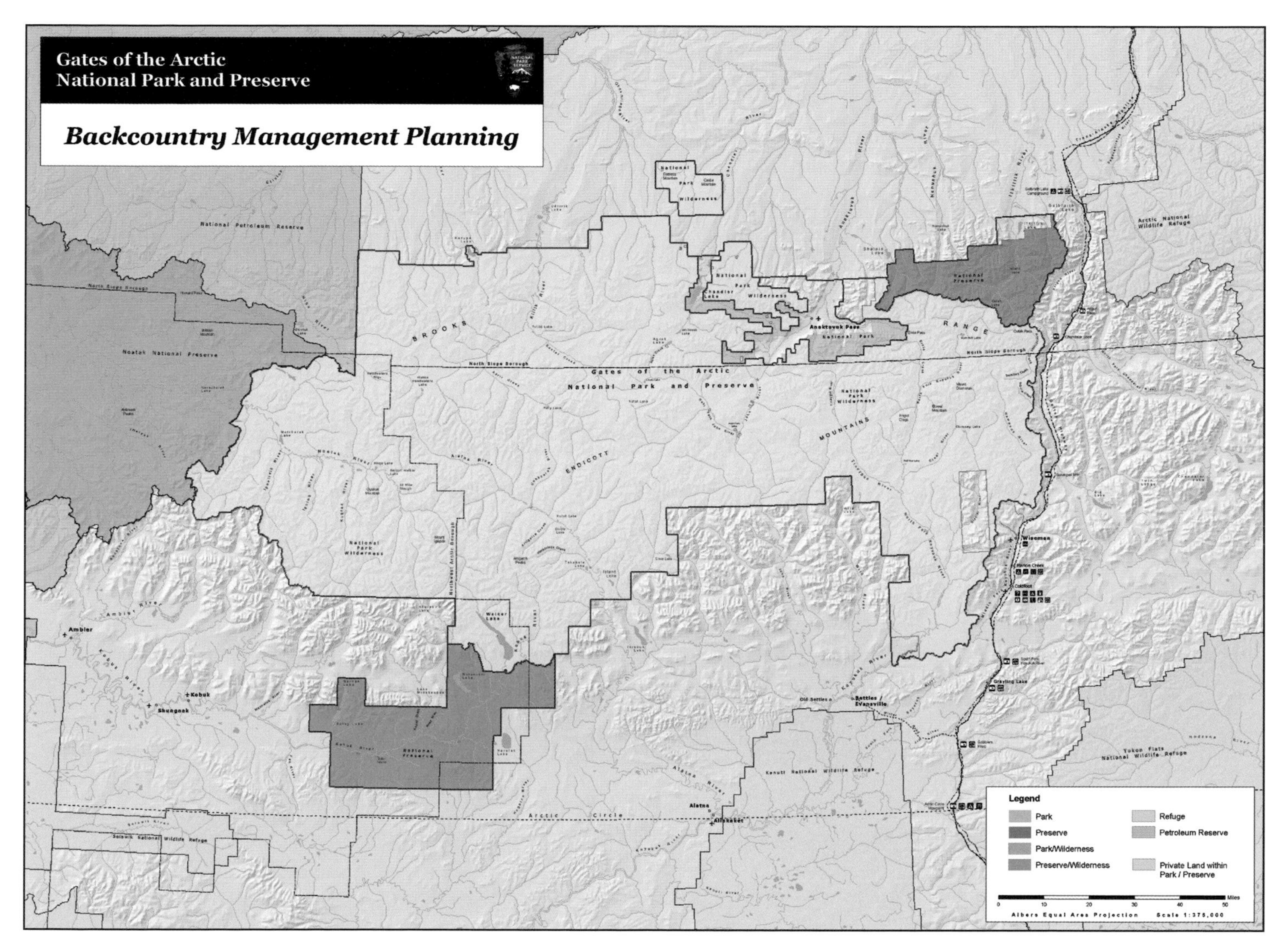

Color coding identifies Gates of the Arctic National Park and Preserve and surrounding land designations. The map helps backcountry management planners develop visitor use and wilderness protection plans for the region.

GIS Application Development in the Alaska Region

National parks in Alaska present unique challenges when it comes to using GIS technology. Most of the parks are remote with limited Internet access. They have relatively small staff size and high turnover, with few if any full-time GIS specialists. These conditions raise the risk of data becoming lost or unavailable, and new personnel face a high learning curve in performing basic GIS functions. Responding to these challenges, the National Park Service has created relatively easy-to-use tools to help novice and infrequent users of GIS make the most of this technology in Alaska. The applications include NPS Theme Manager, AlaskaPak Toolkit, NPS FirePak, and ArcView to Access Link. With a few mouse clicks, for example, NPS Theme Manager lets park staff locate and display more than fifteen hundred available themes, or map layers, for Alaska parks. AlaskaPak Toolkit is a collection of useful GIS tools designed mainly for novice GIS users. NPS FirePak helps users create a standardized data set as well as create GIS maps of wildfire perimeters. ArcView to Access Link uses ESRI ArcView and Microsoft® Access software to set up a live link between GIS maps and computer databases.

TOM WILEY, NPS PHOTO

Caribou grazing at Denali National Park and Preserve.

TOM WILEY, NPS PHOTO

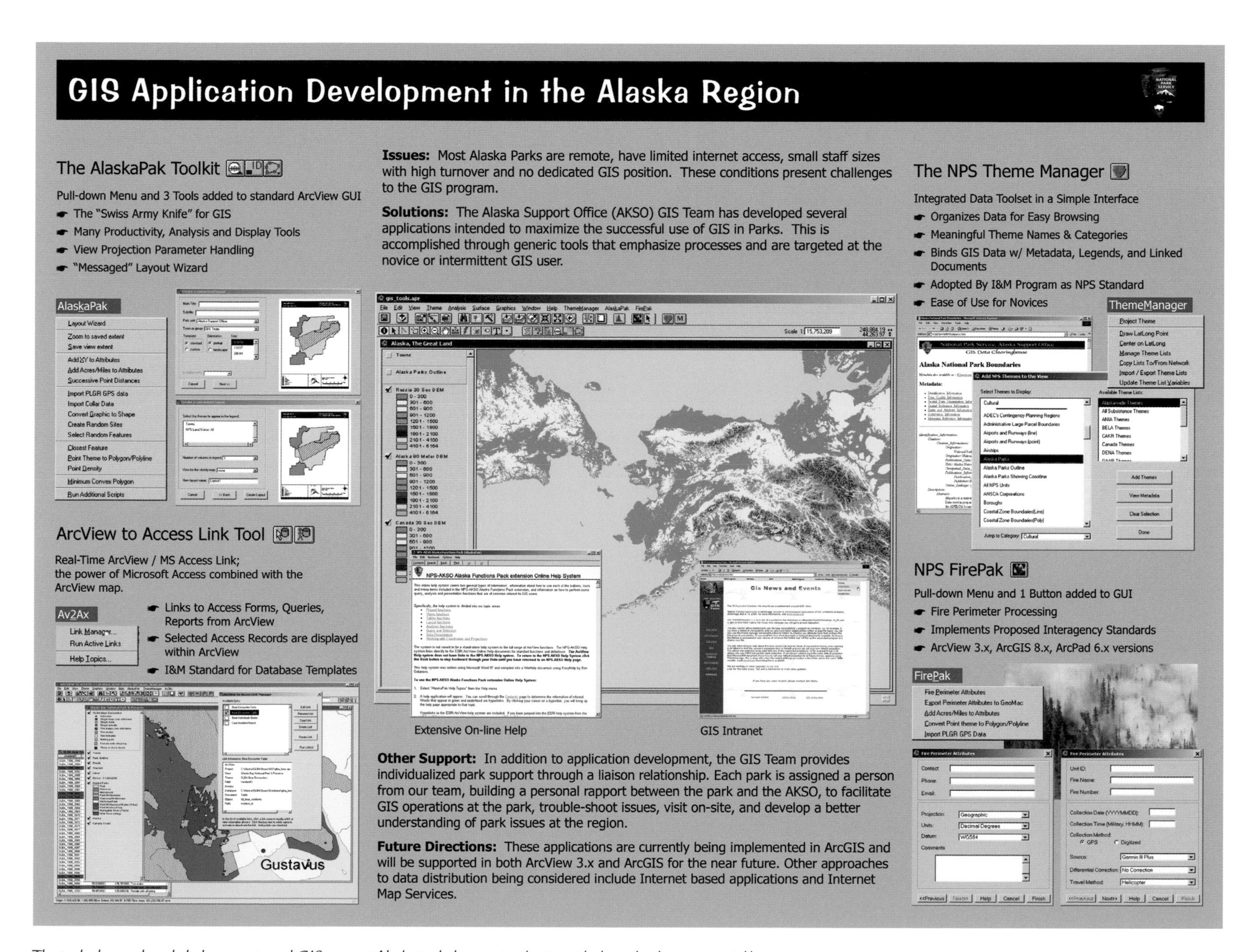

GIS Application Development in the Alaska Region
The AlaskaPak Toolkit
Pull-down Menu and 3 Tools added to standard ArcView GUI
The "Swiss Army Knife" for GIS
Many Productivity, Analysis and Display Tools
View Projection Parameter Handling
"Messaged" Layout Wizard
AlaskaPak
Layout Wizard
Zoom to saved extent
Save view extent
Add XY to Attributes
Add Acres/Miles to Attributes
Successive Point Distances
Import PLGR GPS data
Import Collar Data
Convert Graphic to Shape
Create Random Sites
Select Random Features
Closest Feature
Point Theme to Polygon/Polyline
Point Density
Minimum Convex Polygon
Run Additional Scripts
ArcView to Access Link Tool
Real-Time ArcView / MS Access Link;
the power of Microsoft Access combined with the ArcView map.
Av2Ax
Link Manager...
Run Active Links
Help Topics...
Links to Access Forms, Queries, Reports from ArcView
Selected Access Records are displayed within ArcView
I&M Standard for Database Templates
Gustavus
Issues: Most Alaska Parks are remote, have limited internet access, small staff sizes with high turnover and no dedicated GIS position. These conditions present challenges to the GIS program.
Solutions: The Alaska Support Office (AKSO) GIS Team has developed several applications intended to maximize the successful use of GIS in Parks. This is accomplished through generic tools that emphasize processes and are targeted at the novice or intermittent GIS user.
Alaska, The Great Land
NPS-AKSO Alaska Functions Pack extension Online Help System
Gis News and Events
Extensive On-line Help
GIS Intranet
Other Support: In addition to application development, the GIS Team provides individualized park support through a liaison relationship. Each park is assigned a person from our team, building a personal rapport between the park and the AKSO, to facilitate GIS operations at the park, trouble-shoot issues, visit on-site, and develop a better understanding of park issues at the region.
Future Directions: These applications are currently being implemented in ArcGIS and will be supported in both ArcView 3.x and ArcGIS for the near future. Other approaches to data distribution being considered include Internet based applications and Internet Map Services.
The NPS Theme Manager
Integrated Data Toolset in a Simple Interface
Organizes Data for Easy Browsing
Meaningful Theme Names & Categories
Binds GIS Data w/ Metadata, Legends, and Linked Documents
Adopted By I&M Program as NPS Standard
Ease of Use for Novices
ThemeManager
Project Theme
Draw Lat/Long Point
Center on Lat/Long
Manage Theme Lists
Copy Lists To/From Network
Import / Export Theme Lists
Update Theme List Variables
Alaska National Park Boundaries
Add NPS Themes to the View
Add Themes
View Metadata
Clear Selection
Done
NPS FirePak
Pull-down Menu and 1 Button added to GUI
Fire Perimeter Processing
Implements Proposed Interagency Standards
ArcView 3.x, ArcGIS 8.x, ArcPad 6.x versions
FirePak
Fire Perimeter Attributes
Export Perimeter Attributes to GeoMac
Add Acres/Miles to Attributes
Convert Point theme to Polygon/Polyline
Import PLGR GPS Data

The tools shown above help less experienced GIS users at Alaska parks become productive with the technology more quickly.

Wilderness Expansion 2002

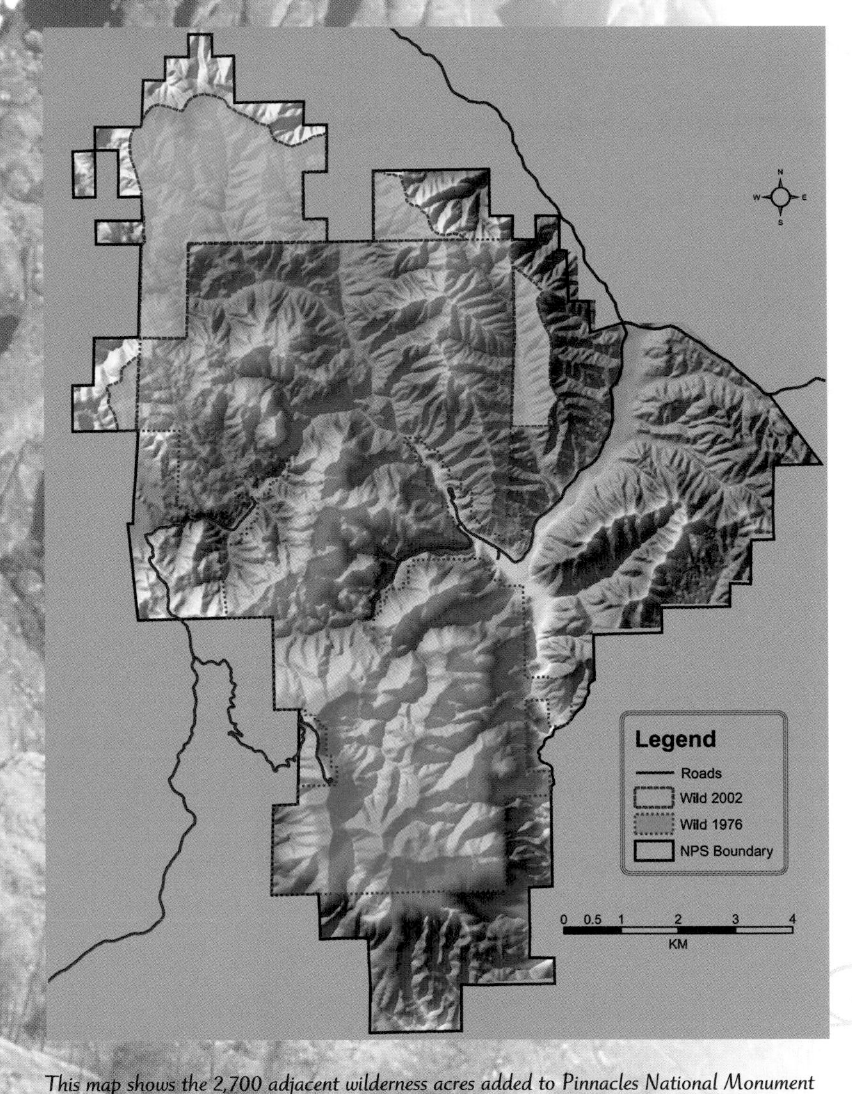

This map shows the 2,700 adjacent wilderness acres added to Pinnacles National Monument in 2002.

Pinnacles National Monument prides itself on preserving 16,000 acres of designated wilderness within a relatively small area in central California. Adding new wilderness requires federal legislation and presidential approval. These are wild places where humans are visitors only. Thoreau's words, "In wildness is the preservation of the world," seem to sum up the National Park Service's philosophy. As the park service and public press for more wilderness to let natural processes function without human influence and to allow more nonmotorized recreation, we also need to make sure that park service projects and activities do not compromise the wilderness designation.

In 2002, the Pinnacles grew by 2,700 acres of wilderness to include adjacent federal Bureau of Land Management land. The monument now encompasses 24,000 acres, including the 16,000 acres of wilderness. The northern part of the park received the largest amount of new wilderness. A GIS map added a new layer using the congressionally designated map and old wilderness areas. Now the park service can easily query the GIS to see where wilderness areas may conflict with projects. Future GIS mapping projects will include a vegetation map showing the monument expansion in 2000. Monument staff also are collecting data on invasive weeds, lichens, amphibians, rodents, moths, butterflies, and birds to improve our understanding of the wilderness. The updated wilderness map boundaries are an important piece of information for resource managers to ensure quality wilderness management and research. The resulting map, available to the public online, also helps park managers understand the kinds of human activity that can take place legally.

Elegant Clarkia wildflowers at Merrion Narrows in Pinnacles National Monument.

At sunset, looking west at the pine and oak trees along a dry creek bed.

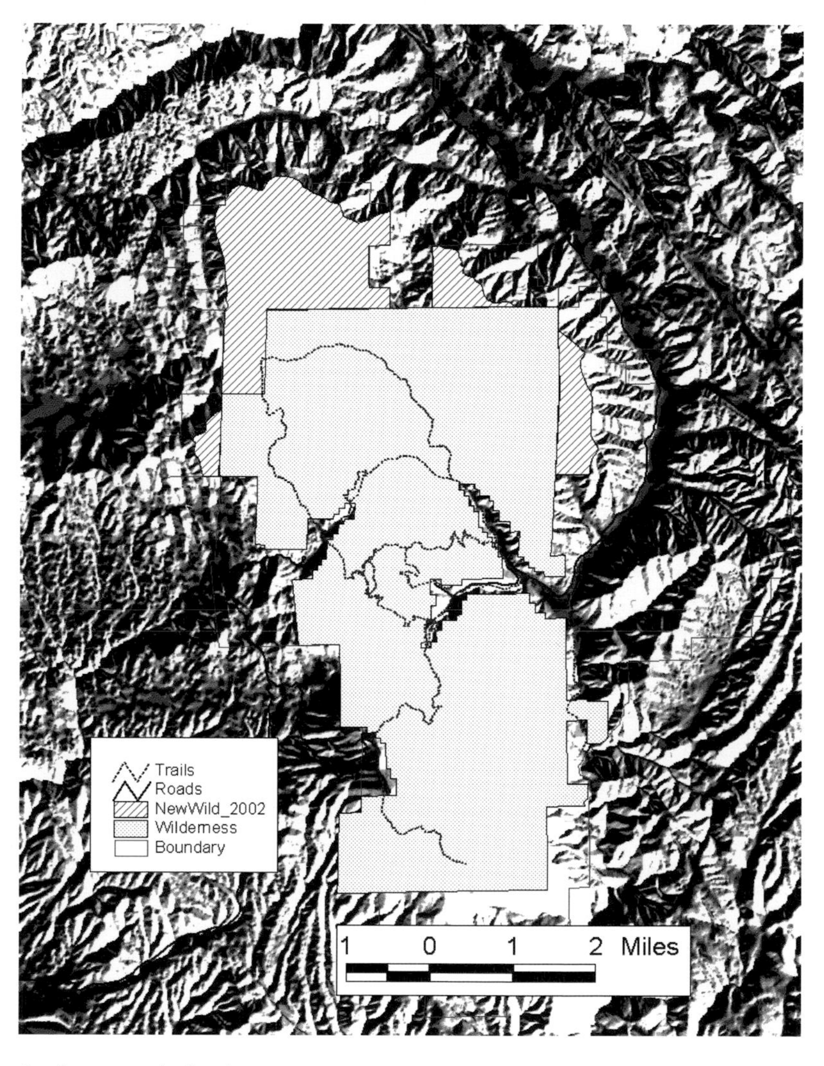

Another way to display the expansion.

Backcountry Inventory

From a distant highway, the majestic peaks of Grand Teton National Park offer a breathtaking view of nature's power and beauty. A closer look reveals the human imprint, from roads and trails to campsites and cellular phone towers. Using GIS technology, the National Park Service has identified the most pristine lands of the region to preserve them for future generations. The effort in Wyoming is part of a larger campaign to identify and preserve wilderness in the Greater Yellowstone Ecosystem. To identify remaining pristine lands in the Tetons, the park service first mapped areas that are not wild. Park rangers and volunteers spent two summers gathering data using GPS technology and digital cameras to document trails, campsites, signs, bridges, and cabins. GIS technology made it possible to analyze layers of data on maps, not only showing trail and campsite locations, for example, but their conditions and need for repairs. When combined with data on visitor use, the GIS maps will model the variation of wildness in the park and help guide preservation strategies across the ecosystem.

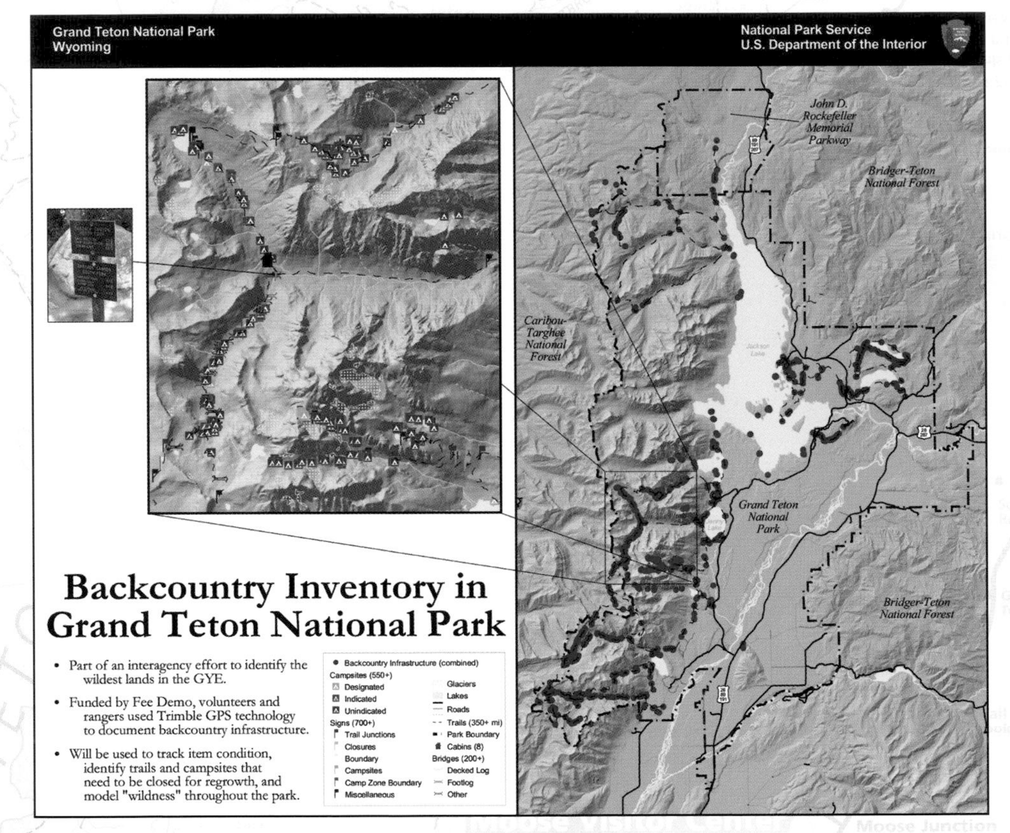

These GIS maps, part of an interagency effort to identify the wildest lands in the Greater Yellowstone Ecosystem, display the backcountry inventory, including park boundary, glaciers, lakes, and trails and campsites that need to be closed for regrowth.

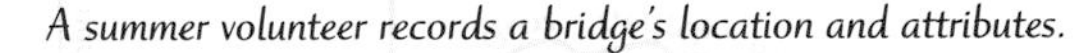

A summer volunteer records a bridge's location and attributes.

A climber on the Exum Ridge of the Grand Teton, the highest of eight peaks above twelve thousand feet elevation in the park.

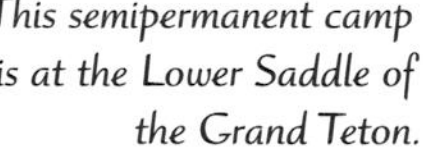

This semipermanent camp is at the Lower Saddle of the Grand Teton.

A sign at the Lower Saddle advises climbers that the fragile alpine tundra requires hundreds of years to grow or recover from trampling; to please step on rocks, not plants; and to camp only on existing bare-ground sites.

THE LOWER SADDLE
FRAGILE ALPINE TUNDRA: THIS VEGETATION
REQUIRES HUNDREDS OF YEARS TO GROW
OR RECOVER FROM TRAMPLING. PLEASE
STEP ON ROCKS, NOT PLANTS. USE THE
OUTLINED PATHS AND CAMP ONLY AT THE
EXISTING BARE GROUND SITES.
PROTECT THE WATER-USE THE TOILET

Hikers trudge the North Fork of Cascade Canyon.

The National Park Service uses this cabin along Berry Creek during backcountry patrols.

The outdoor restroom at the Lower Saddle offers an incomparable view in the park.

Chapter 3

Alive and Well: Park Health and Quality

This semipermanent camp is at the Lower Saddle of the Grand Teton.

THE LOWER SADDLE
FRAGILE ALPINE TUNDRA: THIS VEGETATION REQUIRES HUNDREDS OF YEARS TO GROW OR RECOVER FROM TRAMPLING. PLEASE STEP ON ROCKS. NOT PLANTS. USE THE OUTLINED PATHS AND CAMP ONLY AT THE EXISTING BARE GROUND SITES.
PROTECT THE WATER-USE THE TOILET

A sign at the Lower Saddle advises climbers that the fragile alpine tundra requires hundreds of years to grow or recover from trampling; to please step on rocks, not plants; and to camp only on existing bare-ground sites.

Hikers trudge the North Fork of Cascade Canyon.

The National Park Service uses this cabin along Berry Creek during backcountry patrols.

The outdoor restroom at the Lower Saddle offers an incomparable view in the park.

CHAPTER 3

Alive and Well: Park Health and Quality

The parks are the Nation's pleasure grounds and the Nation's restoring places. . . .
The national parks . . . are an American idea; it is one thing we have that has not been imported.

J. Horace McFarland, president, American Civic Association, 1916

Any visit to a national park raises expectations of what we will experience during our vacation. We may take for granted the scenic views of Yosemite, the clean waters of the Everglades, the multitudes of stars above Denali, or the natural quiet of a backcountry trail in Yellowstone. These natural resources may not have been the reason for the creation of a particular park, but they are fundamental to our expectations. If hazy air causes our eyes to water, if water smells bad and turns orange, if we cannot sit peacefully enjoying the moon and stars, then our parks will be less special, less full of the natural wonder we expect.

Information about the geographic location of places sometimes deals with intangibles and temporary events that are important to the protection of park health. In an age of technology, the National Park Service uses GIS as a tool to help characterize these intangibles and events and guide park managers as they make decisions. We know that air pollution can limit visibility and harm plants, wildlife, and humans. Using GIS technology, we can map the surface patterns of air pollution to help identify which parks are at risk. From a limited set of measurements we can project the extent of the pollution geographically. Even better, we can play make-believe and use sophisticated computer modeling to predict how pollution will move around. More tangible perhaps, but still highly mobile, is our water. To manage our springs, streams, and lakes we need to know where they are and how the water moves. We need to understand water quality and geographic factors that might affect it. Bringing together that information and sharing it on a map is an important management tool. Noise is another threat to park health. Parks cannot avoid all noise intrusions, such as the roar of military aircraft on training flights. Yet working together, we can use GIS mapping to help select air corridors that avoid the most sensitive areas and protect the quietude for park visitors. These examples of GIS projects all share a theme, the careful protection of our parks. As we understand the geographic relationships of threats to our parks, we can manage them in ways that the ranger on patrol or the guard at the gate could not do before GIS.

Puget Sound Basin
Air Pollutant Visualization

The snow-capped icon of the Pacific Northwest rises more than fourteen thousand feet in the Cascades as a dominant testament to nature's power and glory. Visible across the landscape of western Washington State, the active volcano called Mount Rainier contains twenty-six named glaciers, supports hundreds of animal and plant species, and draws nearly two million visitors each year. For all its popularity and beauty, Mount Rainier National Park would like to rid itself of one unwanted guest: air pollution. Nitrous oxides, mainly from motor vehicles, have hurt plants and marine life and degraded scenic vistas for many national parks, including Mount Rainier in the Puget Sound Basin. To educate the public, the National Park Service used GIS to create a three-dimensional display of air pollution flowing through the basin. Working with the University of Washington, the park service made an animated illustration of nitrous oxides sweeping south to Mount Rainier on a typical summer day, dissipating after rush hour. This buildup leads to ozone, which itself hurts the park. The GIS animation illustrated the problem and a solution: reducing automobile emissions that include nitrous oxides and other pollutants.

This sequence of 3-D maps displays the spread of nitrous oxide pollutants through the Puget Sound Basin toward Mount Rainier during the morning rush hour, at midafternoon, and after the evening commute on an average summer day in 2002.

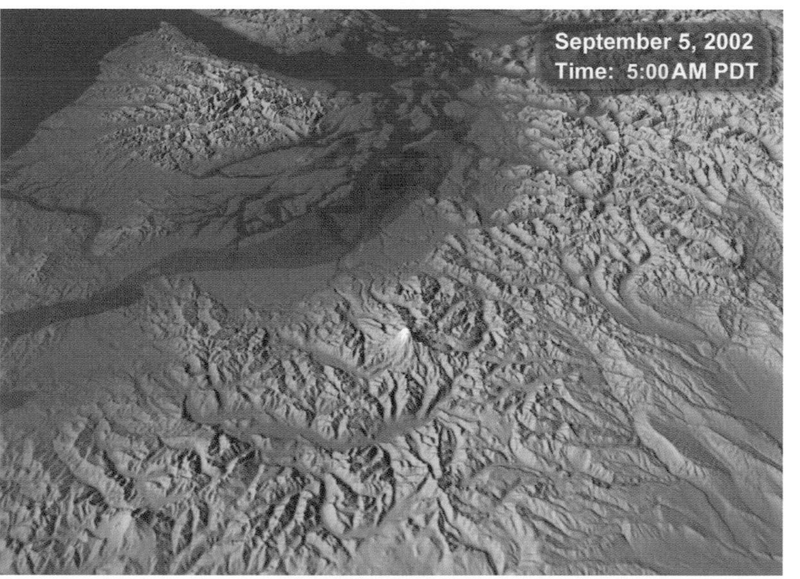

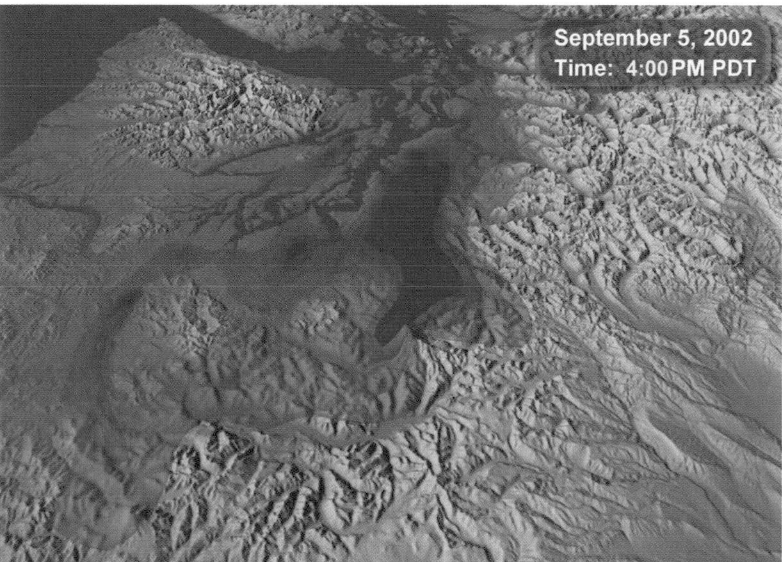

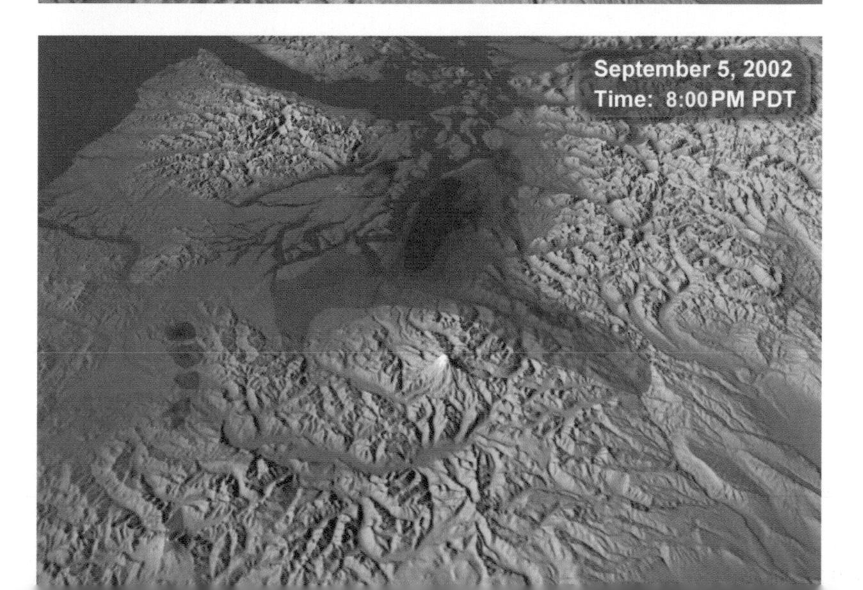

Katmai Water Resources Management Plan

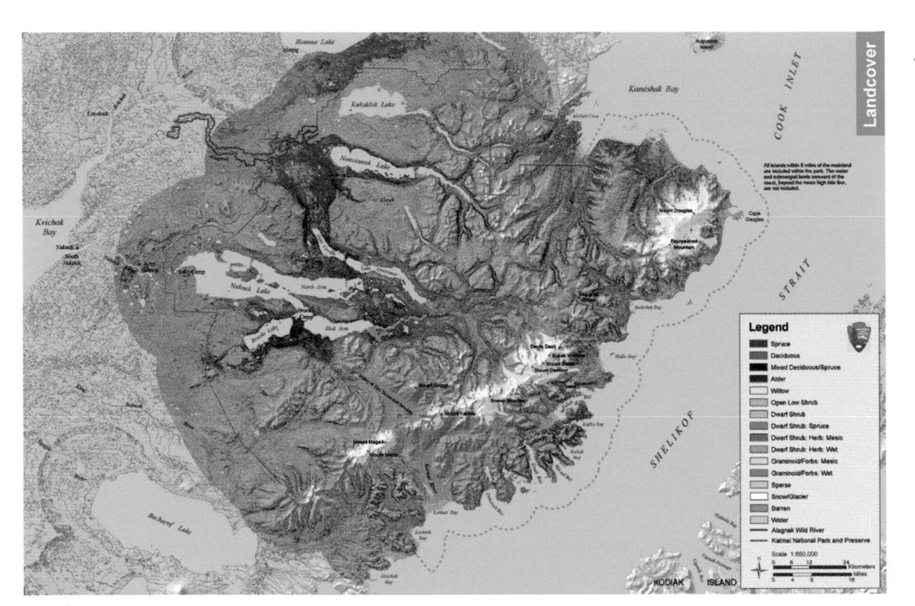

Color coding displays land cover ranging from spruce to barren ground at Katmai National Park and Preserve.

In 1918, Katmai National Monument was created to preserve the famed Valley of Ten Thousand Smokes, where Novarupta Volcano deposited a dramatic 40-square-mile, 100-to-700-foot-deep pyroclastic ash flow. The monument has grown into Katmai National Park and Preserve and is famous for volcanoes, brown bears, fish, and rugged wilderness. It is also the site of the Brooks River National Historic Landmark. The national park and the Alagnak Wild River contain some of the most spectacular waters in the nation, including the largest freshwater lake in the national park system and one of the longest contiguous coastlines at 398 miles. These lakes and streams are remarkable for their color, clarity, and size, as well as their ability to support large numbers of fish. The National Park Service relies on modern technology to safeguard these wonders. The national park is writing a water management plan, the first of its kind in Alaska. The plan includes a series of thirteen GIS maps designed in a clear and informative way to reach a wide audience. The maps depict geology, land ownership, political boundaries, landscapes, coastal resources, water-sampling stations, and other data. The set of maps met the overall goal of a consistent layout, look, and feel. GIS software helped integrate diverse data in the maps. The plan itself will help develop a strategy to manage waters at the 4.1-million-acre park, providing information that managers and policy makers will need to address water resource issues for decades to come.

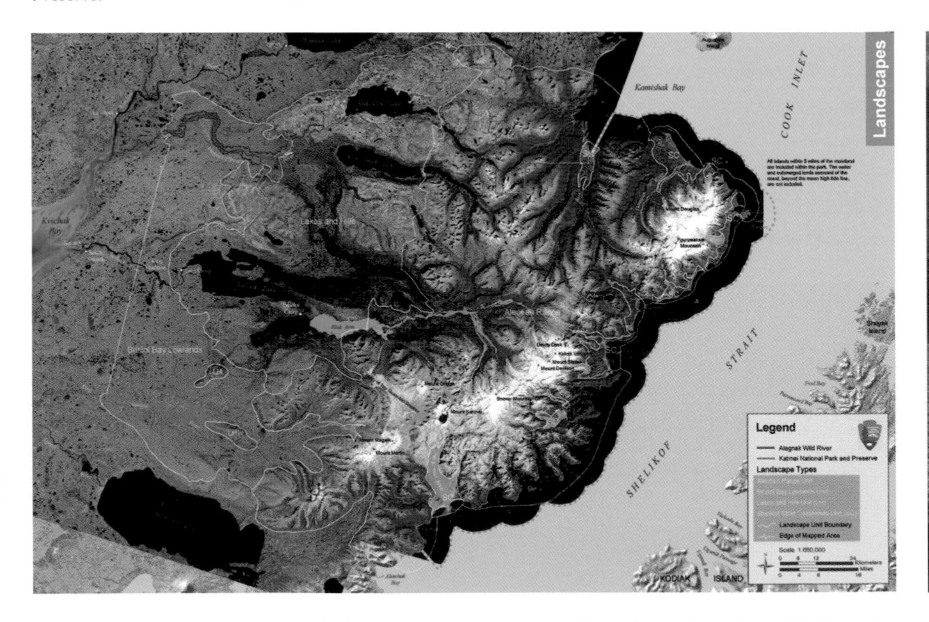

The landscape types at Katmai include the Aleutian Range, Bristol Bay lowlands, lakes and hills, and Shelikov Strait coastlands.

DAVID DURAN, NPS PHOTO

A scene from the Valley of Ten Thousand Smokes.

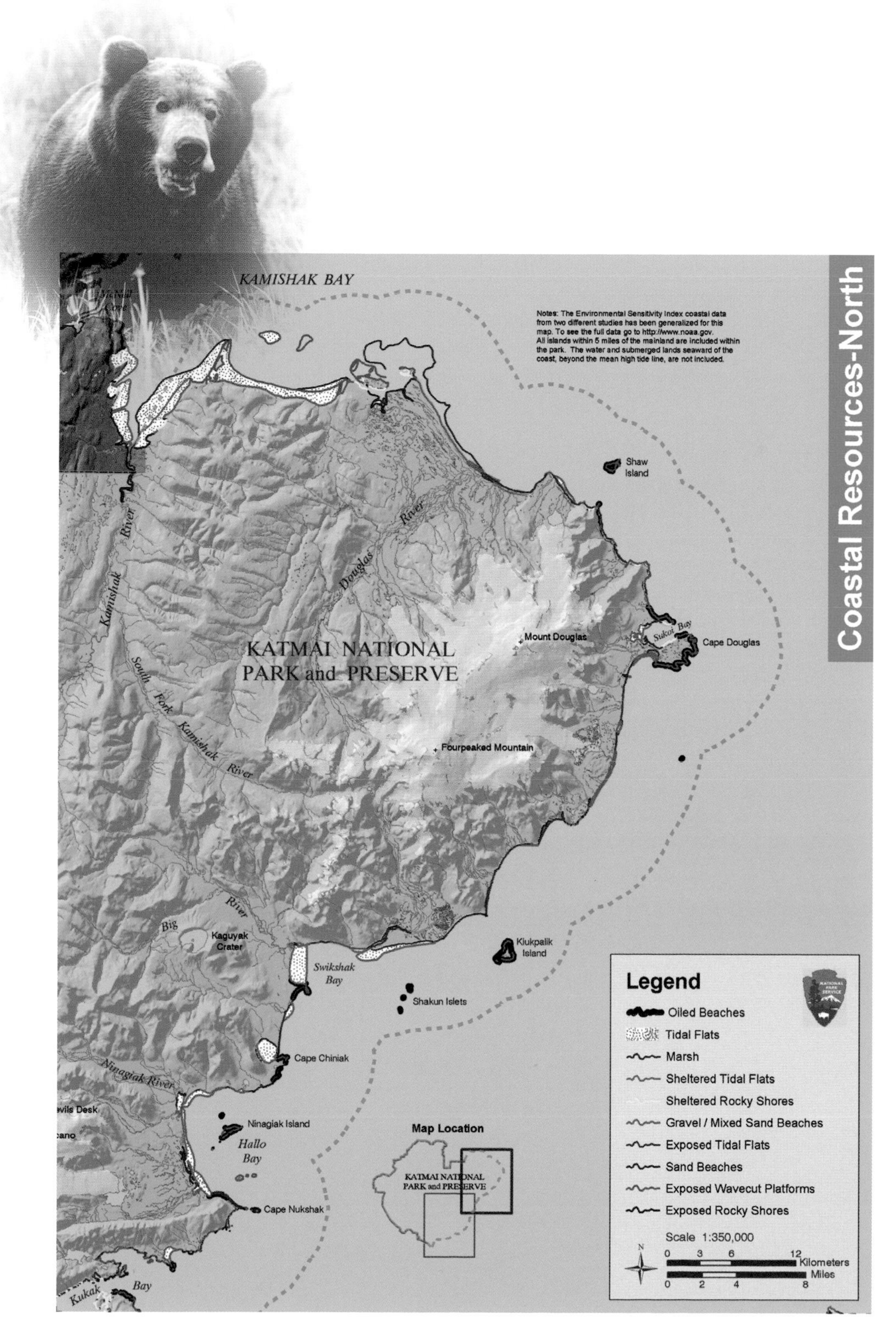

Marshes, tidal flats, and beaches, including those affected by the 1989 Exxon Valdez oil spill, are among the coastal resources identified in this map of Katmai National Park and Preserve.

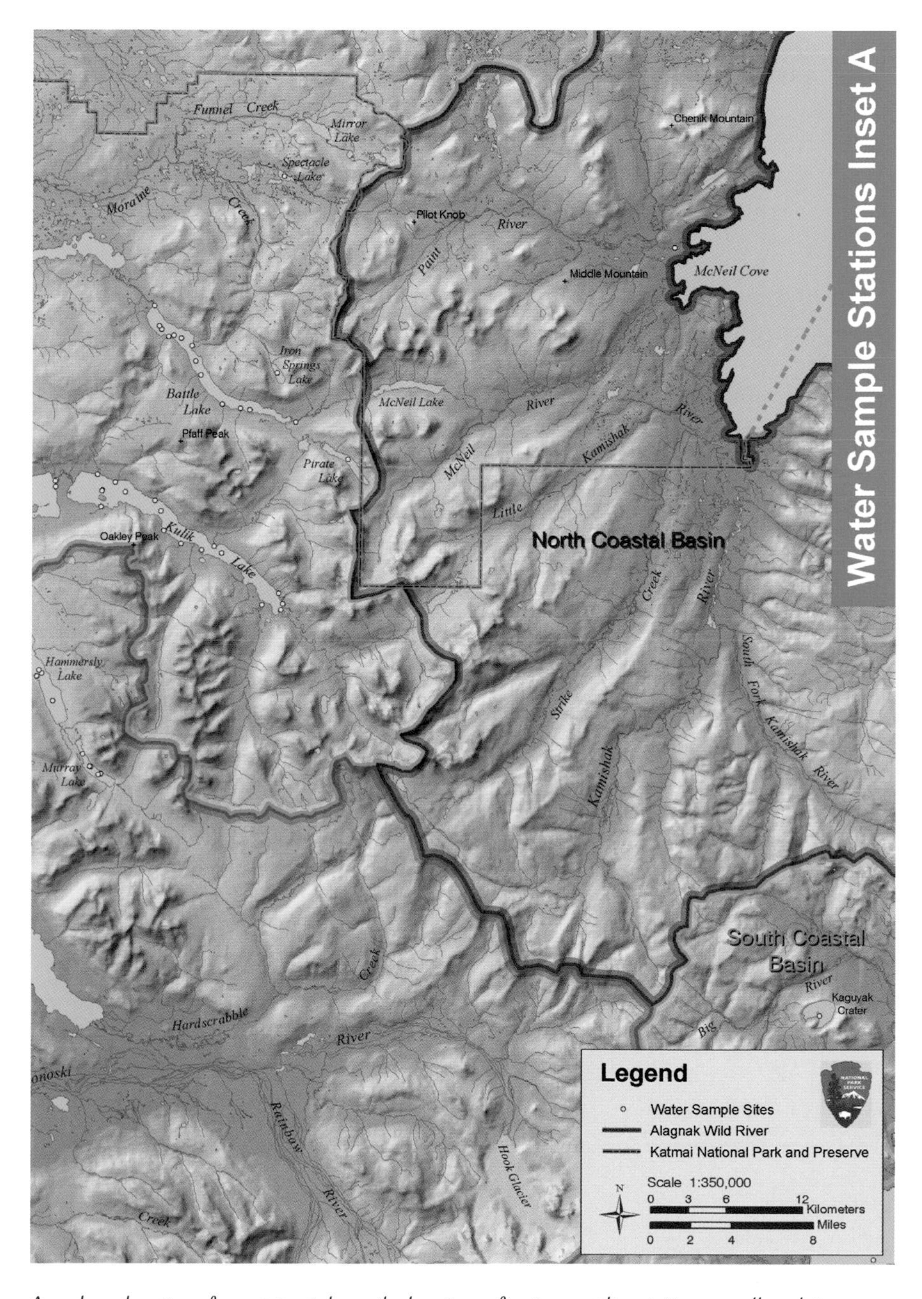

An enlarged section of a map inset shows the locations of water-sampling stations as yellow dots.

Mapping Air Quality Values for the NPS Inventory and Monitoring Program

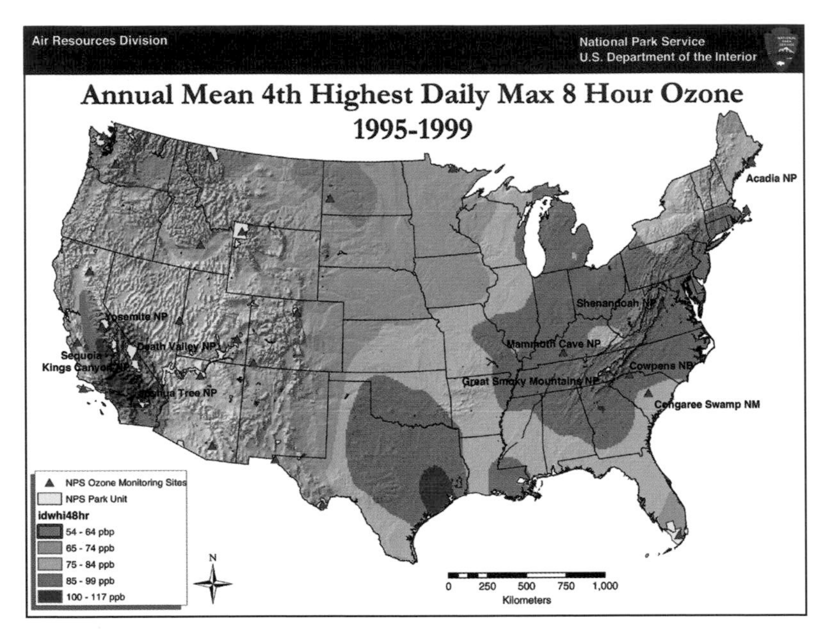

An example of an ozone interpolation map.

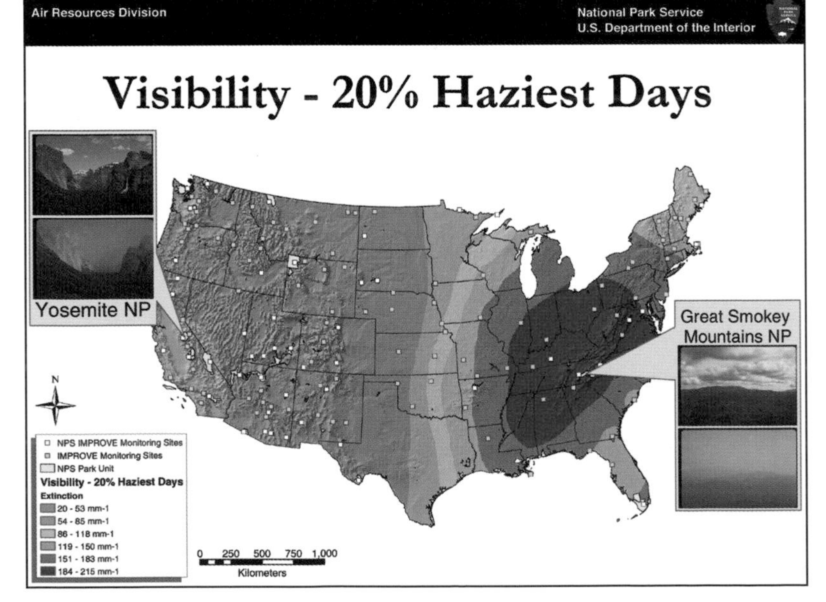

A visibility interpolation map. Examples of the effect of haze are seen in the inset pictures.

We often envision our national parks as havens from civilization where nature prevails and clean air and scenic vistas are the norm. Yet too often we arrive to find smog, haze, and dust filling our lungs and blocking our views. In reality, the air quality in some of our most scenic parks can at times be just as bad as the air we breathe back home. The National Park Service faces the daunting task of measuring air quality and visibility in more than 270 parks. Measuring air quality and visibility can be an intensive and expensive effort. In 2003, only eighty parks had some form of air pollution monitoring in place. The dilemma is how to finish the job in so many parks with such limited funds. The National Park Service turned to GIS in a cooperative mapping effort with the University of Denver. The park service and other federal agencies have collected air quality, visibility, and pollution deposition data in national parks from monitoring stations since the early 1980s. Summarizing the last five years of this data, the park service used GIS software to create color-coded maps of simulated air quality between monitoring stations in the continental United States. The technique showed detailed levels and patterns for ozone and compared them to the national ozone health standard. This resulted in a series of layered GIS maps and tables that estimated air pollution in each of the national parks. These maps now serve as an important tool to determine the need for further monitoring, particularly in parks that might exceed national pollution standards. The park service built an interactive air atlas Web site using ArcIMS® software. Users can view maps, zoom in on a specific park, and query this mini-mapping tool on the Internet. The project gives us a better view of air quality in our national parks in combination with other GIS map layers, such as trends in population and pollution, the presence of sensitive species, and sources of emissions.

A clear-day view of the air quality monitoring station at Turtleback Dome and Yosemite Valley, Yosemite National Park.

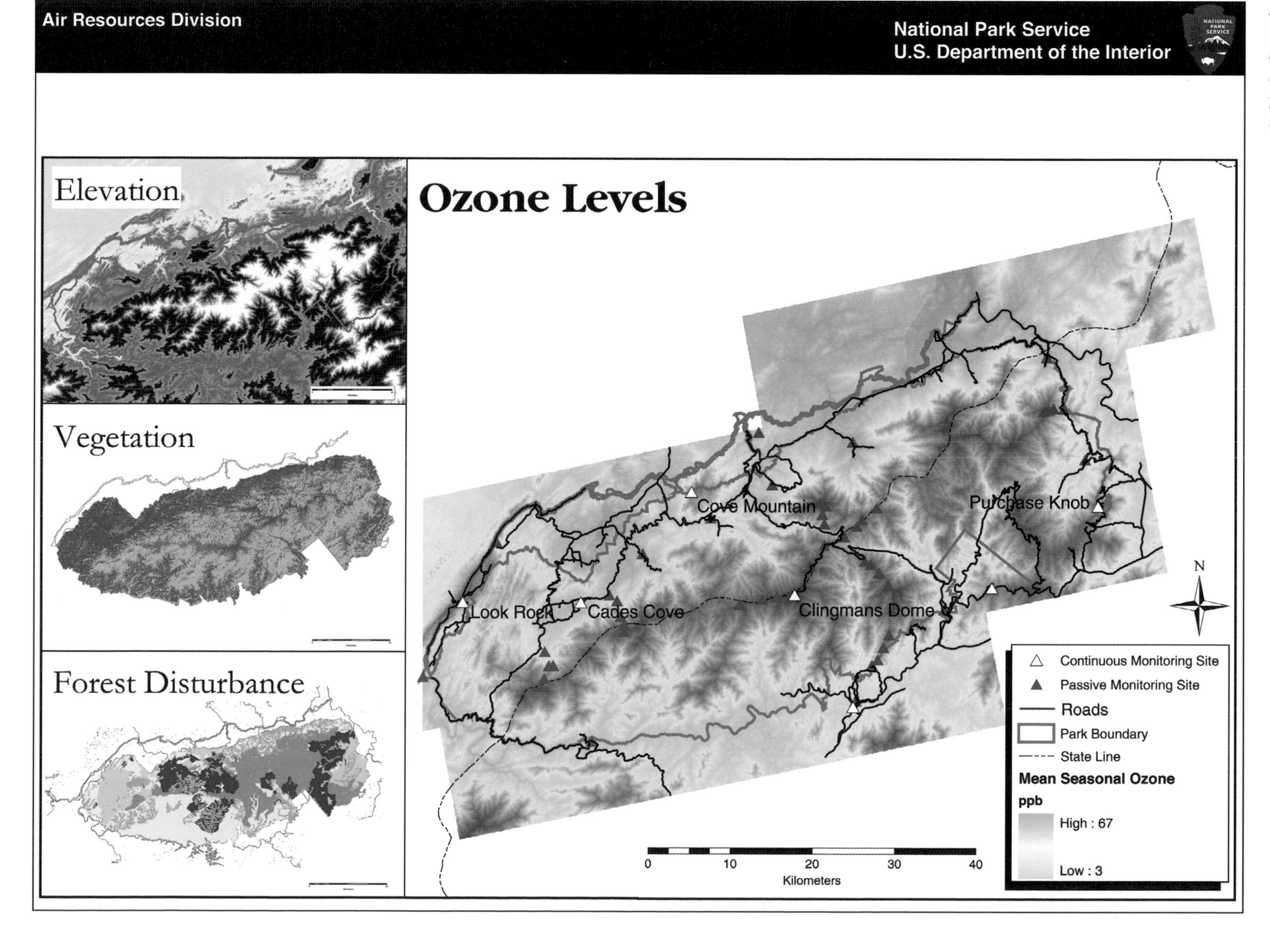

A study in Great Smoky Mountains National Park showed it is possible to display additional detail about ozone concentrations. The inset maps at the left (elevation, vegetation, and forest disturbance) are a small set of GIS layers that can be used with the ozone layer.

Colonial National Historical Park Elevation

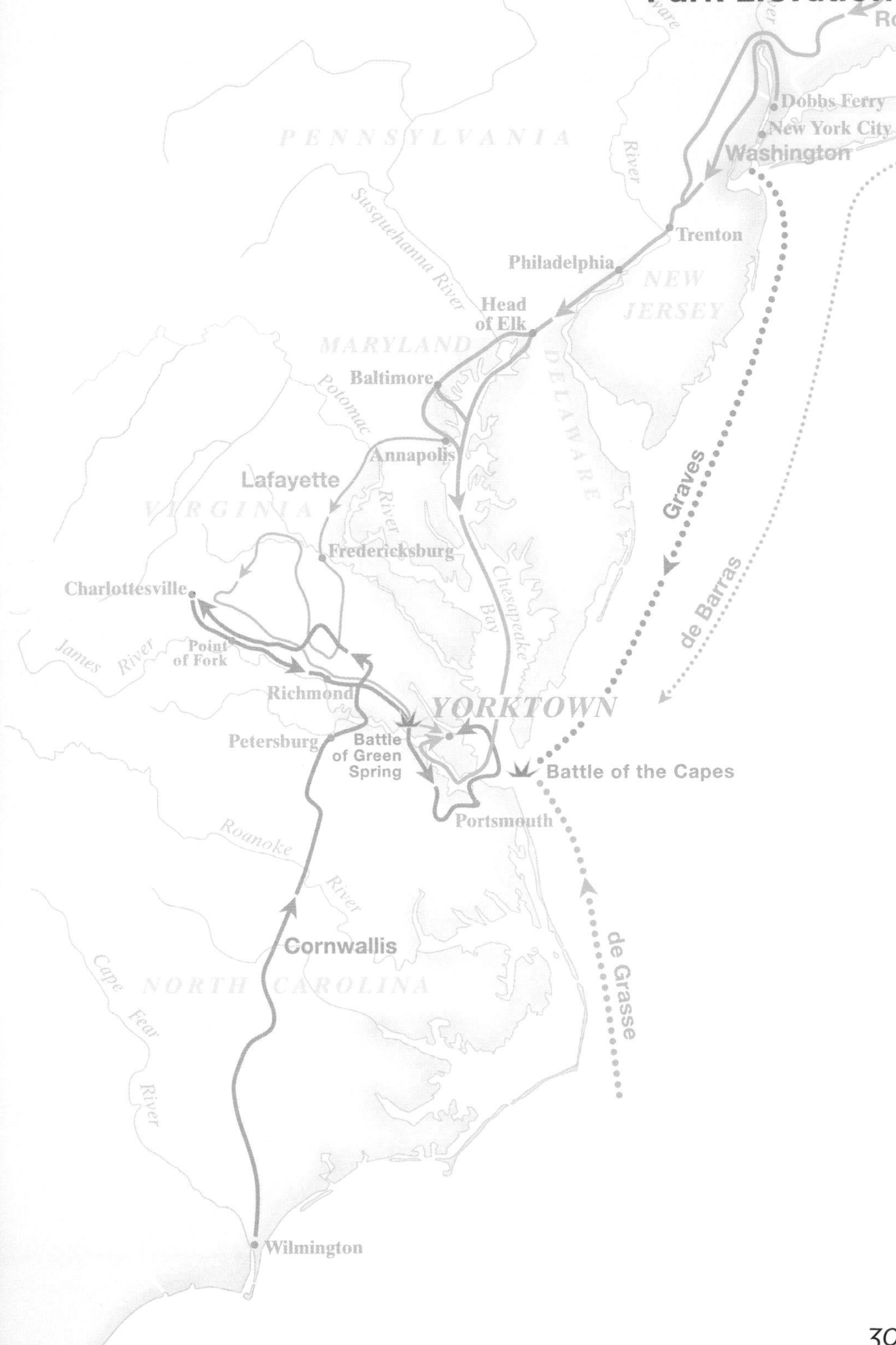

The tidewaters of Virginia include hundreds of wetlands, streams and springs, and drainage systems, some of them within Colonial National Historical Park. Situated on the Virginia Peninsula, the park administers Jamestown, the first permanent English settlement in North America, jointly with the Association for the Preservation of Virginia Antiquities. The park also administers Yorktown Battlefield, site of the last major battle of the Revolutionary War in 1781. The variety of natural resources in addition to wetlands includes forest and fields as well as rare birds, plants, and wildlife. GIS mapping of the elevation data shows that wetlands cover a quarter of the more than twenty-four hundred acres of parkland. The wetlands include more than thirty-two miles of river shoreline and fifty-four miles of seasonal and year-round streams. GIS technology helped create water models with elevation data to illustrate the need to improve and protect these park resources.

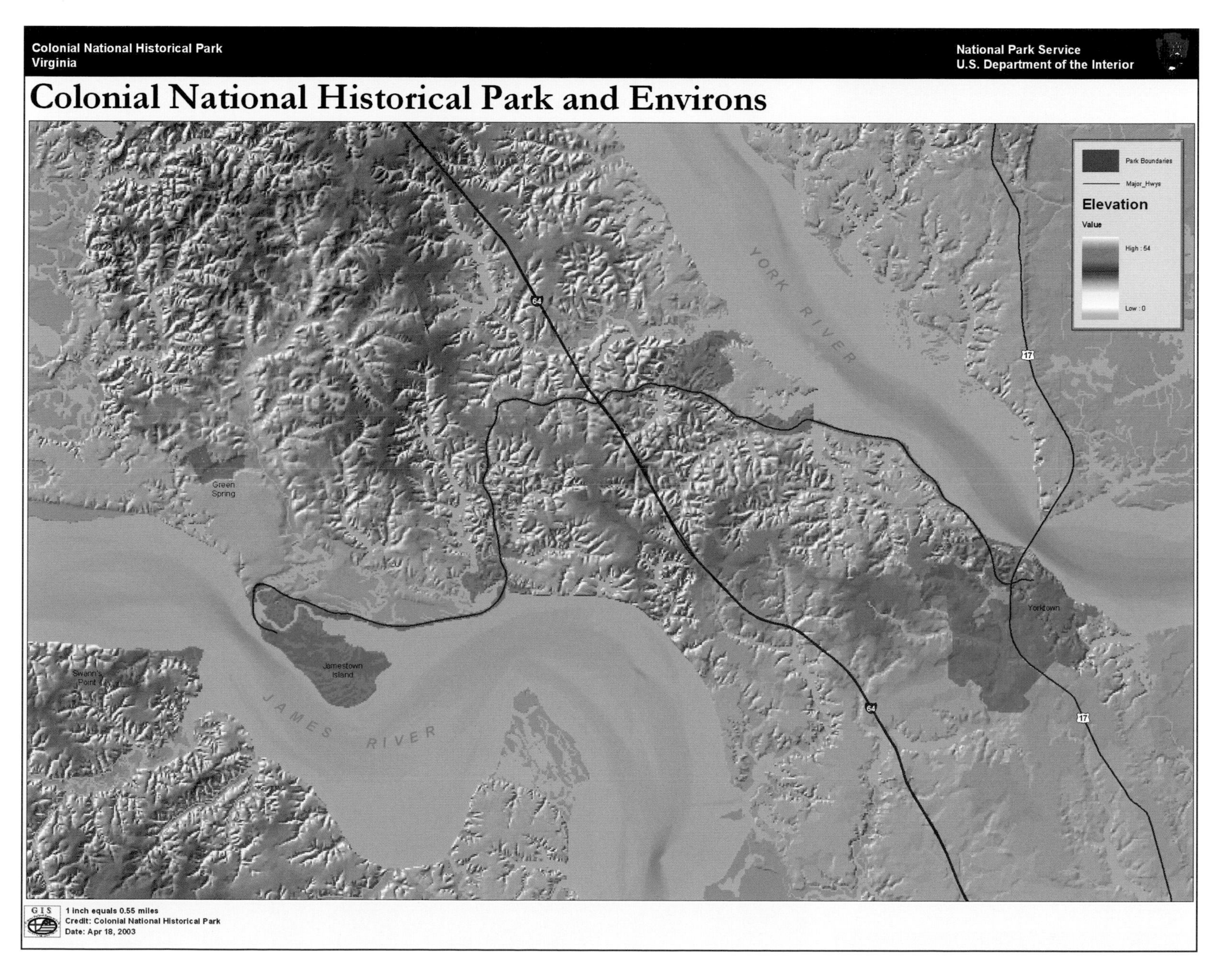

The National Park Service uses elevation maps like this in its efforts to safeguard the park's valuable wetlands.

The U.S. Air Force and National Park Service Western Pacific Regional Sourcebook

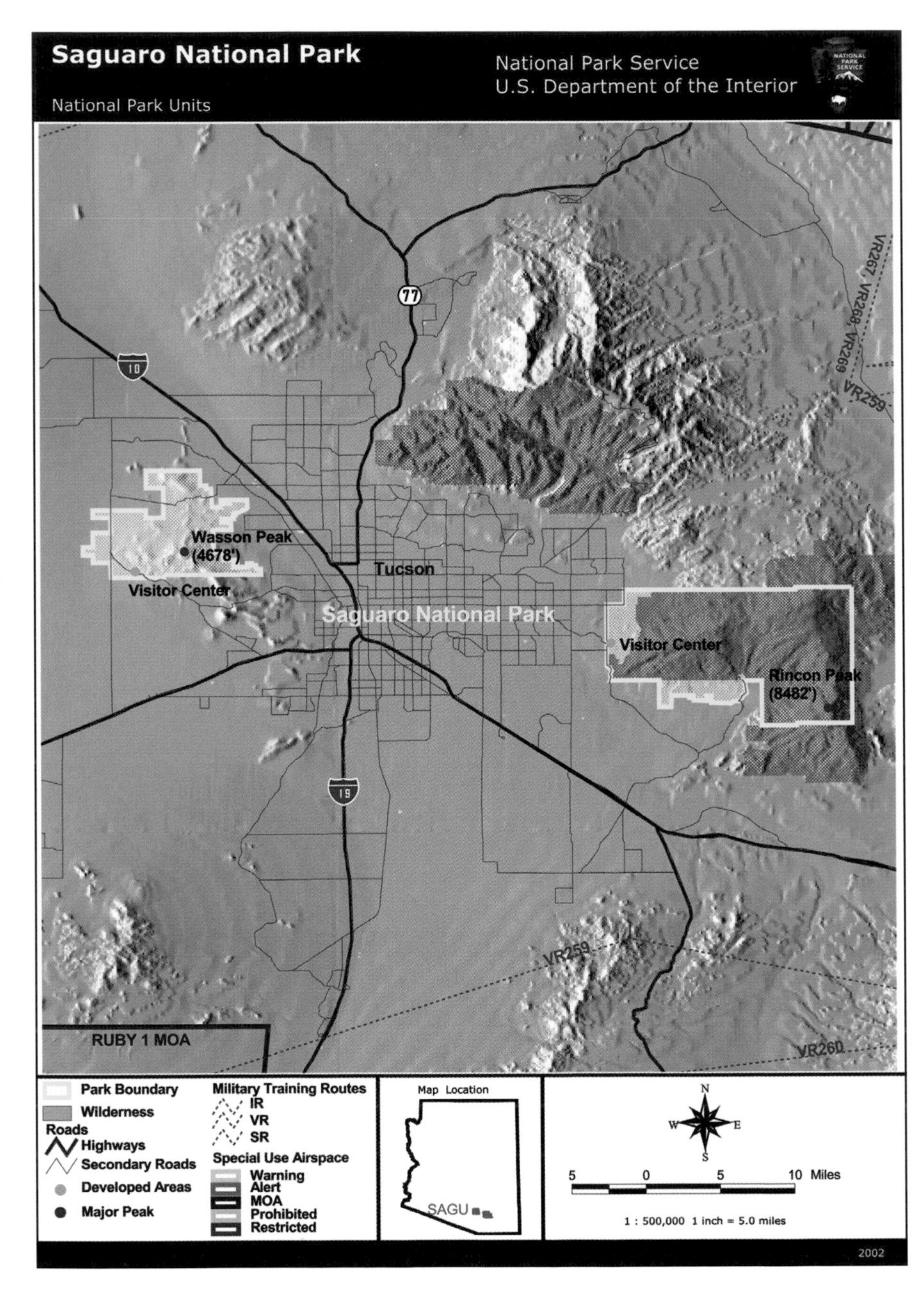

U.S. Air Force training routes and "special use" airspace boundaries near Saguaro National Park.

The U.S. Air Force and National Park Service share a fundamental mission: to ensure that the America we know today is the same nation we pass on to our children and grandchildren. The Air Force must test equipment and train its members to defend our country. The National Park Service must serve, satisfy, and educate visitors while safeguarding wildlife and natural, cultural, and historic resources. Serving both needs at once can be complex and challenging when some 150 of the 388 park units sit under aircraft training routes and other military airspace. Population growth and development around air bases threaten to leave less airspace for military training. Similarly, aircraft noise has intruded on national parks that are set aside as crown jewels to look and sound today and tomorrow as they did generations ago. In a partnership to build mutual understanding and communication, the two agencies created a sourcebook that includes information about the purpose and value of each national park and Air Force base in the region. The book includes GIS maps that cross-reference park boundaries, major geographic features, and visitor services with military airspace and the military uses of that airspace. While some noise intrusion remains inevitable in our national parks, GIS technology has helped the government manage and even avoid these conflicts.

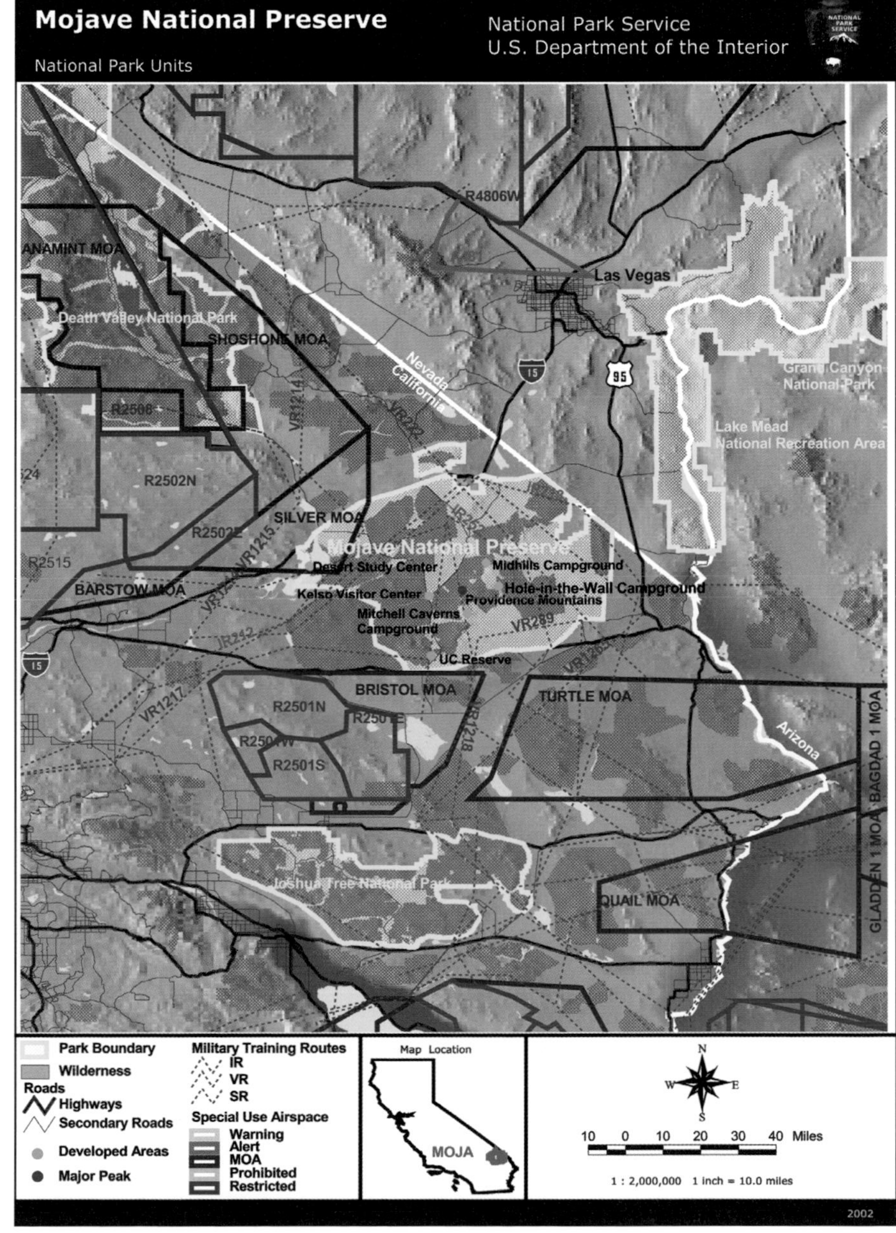

Mojave National Preserve in California (in the center), along with Joshua Tree, Death Valley, and Grand Canyon national parks and Lake Mead National Recreation Area in relation to military aircraft training routes and airspace designated for special uses.

Chapter 4

Without a Trace: Human Impacts

If I can help document . . . change and thereby highlight the effects of our collective actions on the world and particularly on the invertebrate world, a world rarely noticed by humans but comprising over 90% of the species found on Earth, my retirement will be meaningful to me.

Rich Brady, volunteer, Rocky Mountain National Park
From Natural Resource Year in Review, 2002

People lived in places destined to become national parks long before cartographers began mapping their boundaries in the nineteenth century. In some cases, oral histories and traditions place people in these locations since time immemorial. Their lives intertwined inexorably with the environment. They drew nourishment from fishing, hunting, farming, and food gathering, and found spiritual meaning in the sun, moon, animals, plants, and changing seasons around them.

In modern times, many lands once used for homesteading, ranching, logging, mining, and commercial fishing now make up our national parks. Some of these uses scarred the landscape, while others left barely a footprint. In each case, the National Park Service strives to understand and interpret the past while managing for the present and future.

GIS helps us understand the complex and related ways that we use the landscape. The National Park Service uses GIS technology to explain these relationships. In American Samoa, GIS mapping helped outline the different uses of the land, protecting the rain forest and showing residents where they could grow food. In Alaska, GIS helped resolve an important court case regarding mining claims. The technology also helped create the first map designed to better understand and manage subsistence hunting on the Alaska Peninsula. And the federal government uses GIS software to include elders' traditional knowledge of fishing in management plans.

The National Park Service seeks to understand and respect enduring human attitudes about parks. People use these special places for everything from religious ceremonies and recreation to sport and subsistence hunting and fishing. GIS reveals and documents our relationship with parks and helps us all become better stewards of the land.

Mapping Subsistence Agriculture in the National Park of American Samoa

Rory West Jr. takes a GPS position with a Precision Lightweight GPS Receiver along the edge of a taro field on Ta'u Island.

Collaboration between the National Park Service and several Samoan villages lets farmers grow traditional crops within a federally managed preserve. The National Park of American Samoa is unique in that the federal government does not own the land but leases it from several villages for a period of fifty years. The lease agreement protects the integrity of the rain forest, archaeological and cultural resources, and the traditional way of life called fa'asamoa. The matai system of lawful, chiefly authority is one of the most important components of fa'asamoa. For centuries, the matai system has acted to enforce fa'asamoa over the lands and waters now within the national park. The objectives of the national park and fa'asamoa reinforce each other. The National Park Service manages lands and reefs in the park while villages keep their subsistence farming rights. This cooperation extended to the use of GIS technology to map and classify 232 acres within the 9,355-acre park for subsistence farming. Under park lease provisions, native American Samoans can continue subsistence farming on land they have cultivated or fallowed within the last fifteen years, using traditional tools and methods, while clearing and cultivation is prohibited in protected forest. Subsistence agriculture typically includes maintaining small plots of land for the cultivation of traditional Polynesian crops such as bananas, taro, breadfruit, and coconuts. ArcGIS® software integrated data from global positioning systems, remotely sensed imagery such as aerial photographs, and field observations and sketches by botanists and Samoan subsistence farmers. The project overcame the challenges of cloud cover, the rough terrain of a park cloaked in dense tropical rain forest, and the need for cultural sensitivity in an area of communally owned land. GIS maps served as an important tool for cross-cultural communication and to protect America's southernmost rain forest and traditional subsistence farming in the park.

Aerial photograph shows the agriculture of banana and coconut trees next to a freshly planted taro field on Ofu Island.

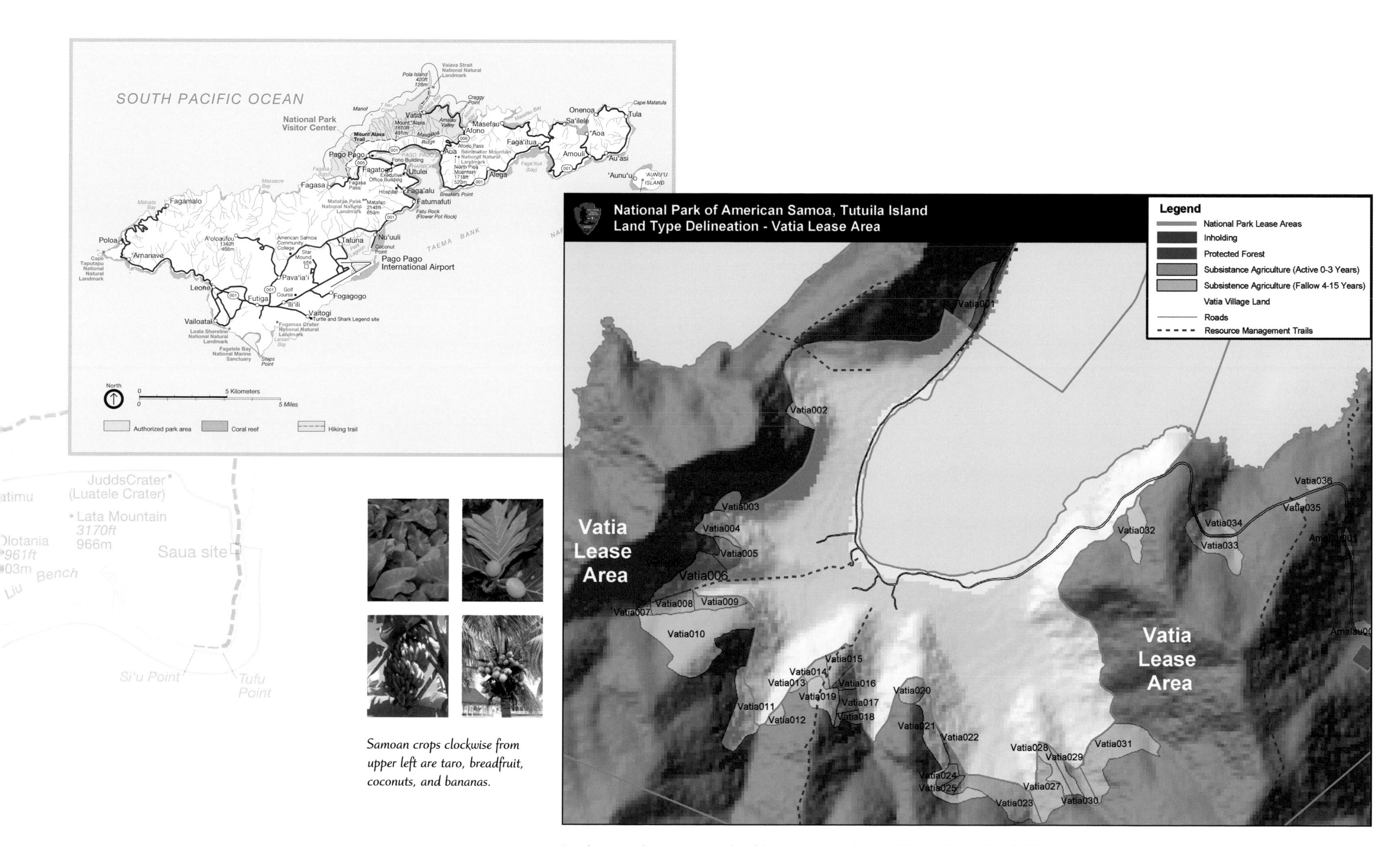

Samoan crops clockwise from upper left are taro, breadfruit, coconuts, and bananas.

Land uses in relation to national park lease areas in and around Vatia village on Tutuila Island.

Kantishna Hills Mining Court Cases, Denali National Park and Preserve

GIS is helping the National Park Service show the extent of activity on mining claims inside Denali National Park and Preserve in Alaska. The 1980 expansion of McKinley National Park that created the park and preserve included the Kantishna Mining District, an area of placer and lode mining claims, mostly for gold. The park service has been buying these claims for preservation whenever possible, sometimes ending up in court over the issue of appraised fair market value. In these disputes, the park has used GIS maps based on data collected from aerial photographs and other technology to explain key issues in court. The maps are made of several components. Using GIS software, a private contractor mapped the mining claims from aerial photographs and ground control surveys with two-foot contours in a process called photogrammetry. The contractor created three-dimensional topographic maps and used this data to produce a digital photomosaic, then overlaid the data on the mosaic. Aerial photographs taken at different phases of the mining showed details of the mining activity. In one instance, GIS maps showed the extent of mining on the upper reaches of Caribou Creek in the preserve. Geologists, mining claim appraisers, engineers, and other witnesses used the exhibits to show that one plaintiff had mined more gold than stated. As a result, a judge valued the mining claims at about seven million dollars less than sought by the plaintiff.

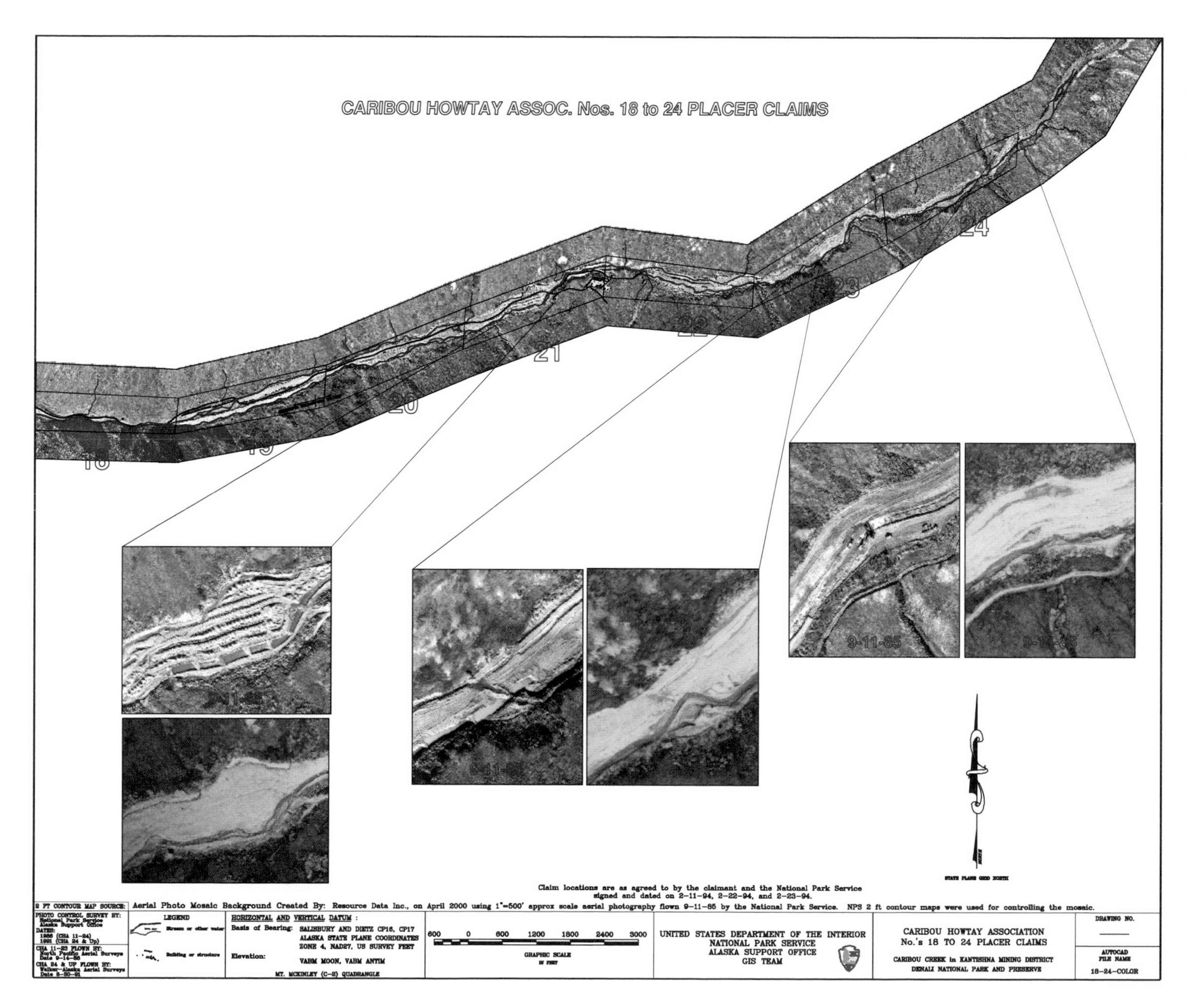

This GIS map is an example of courtroom exhibits used to show the extent of mining in the upper reaches of Caribou Creek.

Subsistence Hunting in Aniakchak National Monument and Preserve

After the Aniakchak caldera formed, an eventual breech in its wall caused a massive flood, creating the great cleft through the caldera wall now known as The Gates. This allows the Aniakchak River to begin its tumultuous twenty-seven-mile course southeastward from Surprise Lake, shown here, to the Pacific Ocean.

A valuable resource, salmon account for nearly 50 percent of the subsistence foods used by residents of communities near the monument and preserve.

Midway down the wild and roadless Alaska Peninsula sits one of the nation's most fascinating volcanic features. Aniakchak is a 6-mile-wide, 2,000-foot-deep caldera, a volcanic feature formed by the collapse of a 7,000-foot mountain. Aniakchak's origin is uncertain but the caldera probably resulted from a horizontal explosion of volcanic gas, dust, superheated steam, and pyroclastic material from a side vent of the volcano. A series of smaller eruptions followed, the latest in 1931. The base of the volcano has a circumference of about 100 miles. The Aniakchak River passes through the east wall at a point known as The Gates, where erosion helped form a narrow canyon between cliffs 2,000 feet high. To the east, rugged bays and inlets of the Pacific coast and offshore islands provide habitat for sea mammals and birds.

In creating Aniakchak National Monument in 1978, Congress recognized the unique geological significance of the caldera. In 1980, Congress added the preserve designation and also acknowledged the wildlife and recreational values of the Aniakchak River by designating it a wild river within the National Wild and Scenic Rivers System. The monument and preserve also contains other important resources, including waterfowl, migratory birds, and sea mammals. For thousands of years, Alaska natives have relied on fish, wildlife, wild plants, and other natural resources to provide food, shelter, clothing, transportation, handicrafts, and trade. Today, many rural Alaska native and nonnative people continue to depend on the land and waters for their survival. For many, the ability to continue these subsistence activities is integrally tied to cultural preservation. The federal government has managed subsistence hunting and trapping on its lands since 1990 and at fisheries in federal waters in Alaska since 1999. The federal government began making GIS subsistence hunt maps on an annual basis in 1999, beginning at Wrangell-St. Elias National Park and Preserve with species-specific hunt maps. These maps required seasonal updates because of changes in land-status data and regulations. The map of Aniakchak National Monument and Preserve is a first for the Alaska Peninsula area. The National Park Service and other federal agencies understand the importance of subsistence hunting for those who depend on it and have used GIS mapping to show where and what people can fish, hunt, and trap under federal subsistence regulations.

Aniakchak National Monument & Preserve
Federal Public Lands Under Subsistence Hunting Regulations

Alaska Region
National Park Service
U. S. Department of the Interior

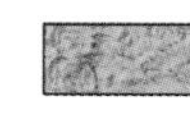
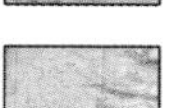

Preserve
Monument
This is not a land status map. This map depicts lands treated as Federal Public Lands for the purpose of subsistence hunting under the Federal Subsistence Management Program Regulations.

Federal Public Land

Non-Federal Public Land

Definition of Federal Public Lands under 50 CFR 100 (Federal Subsistence Management Regulations):

Public lands or public land, or Federal public lands or Federal public land, means lands situated in the State of Alaska which are Federal lands, except:

1) land selections of the State of Alaska which have been tentatively approved or validly selected under the Alaska Statehood Act and lands which have been confirmed to, validly selected by, or granted to the Territory of Alaska under any other provision of Federal law,

2) land selections of a Native Corporation made under the Alaska Native Claims Settlement Act which have not been conveyed to a Native Corporation, unless any such selection is determined to be invalid or is relinquished; and,

3) lands referred to in Section 19(b) of the Alaska Native Claims Settlement Act.

4) however, until conveyed or interim conveyed, all Federal lands within the boundaries of any unit of the National Park System, National Wildlife Refuge System, National Wild and Scenic Rivers Systems, National Forest Monument, National Recreation Area, National Conservation Area, new National Forest of forest addition shall be treated as public lands for the purposes of these regulations.

Explanation of Datasets:

USGS 1:250,000 series topographic quadrangles combined with a 90 meter shaded relief digital elevation model (DEM) provide the background for this map.

The Federal Public Lands within Aniakchak was defined by using a National Park Service land acquisition dataset. The NPS land status mapping is based primarily on Bureau of Land Management (BLM) land status records and plats. These include, but are not limited to, Interim Conveyance and Tentative Approval documents, Patents, Master Title Plats (MTP's), Historical Indexes (HI's), US surveys and US mineral surveys and their respective field notes and Land Information Systems. Subdivision plats and deeds are also used. Mapping is primarily at 1:63,360 scale. The last update was in March 2003.

Federal Public Lands outside of Aniakchak was defined by using a State of Alaska, Department of Natural Resouces (ADNR), generalized land ownership dataset. This GIS coverage uses data extracted from BLM's records, stored in the Alaska Land Information System (ALIS) on January 2, 2001; and ADNR's land records stored in the Land Administration System (LAS) on January 10, 2001. The data is mapped at the section level. No parcel level mapping is included in this dataset. Allottments are portrayed as a dot indicating that one or more allotments are contained within that section.

National Park Service
Alaska Support Office
GIS Team

2525 Gambell Street
Anchorage, Alaska 99503
907-257-2690

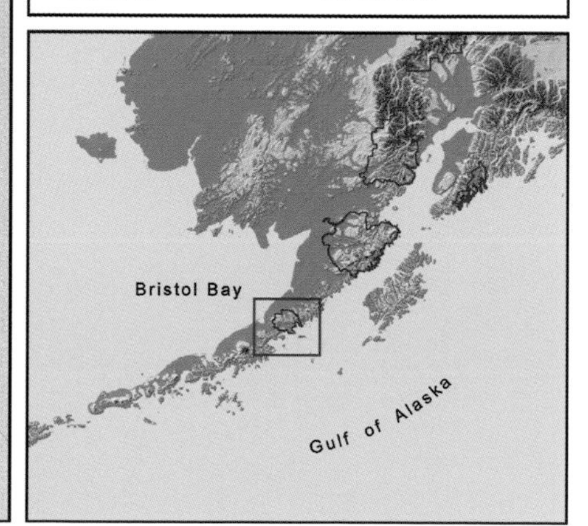
Bristol Bay
Gulf of Alaska

Red lines mark the boundaries of the national monument and preserve. Green areas show where people can hunt, fish, and trap under federal subsistence regulations. Inset identifies these lands on the Alaska Peninsula.

Chapter 5

High-Tech Digs: Archaeology

As we Americans celebrate our diversity, so we must affirm our unity if we are to remain the "one nation" to which we pledge allegiance. Such great national symbols and meccas as the Liberty Bell, the battlefields on which our independence was won and our union preserved, the Lincoln Memorial, the Statue of Liberty, the Grand Canyon, Yellowstone, Yosemite, and numerous other treasures of our national park system belong to all of us, both legally and spiritually. These tangible evidences of our cultural and natural heritage help make us all Americans.

Edwin C. Bearss, National Park Service chief historian, 1981–1994

Archaeologists have always used maps to document and share information about ancient ruins, buried artifacts, and other sites. The use of computers and GIS in archaeology has transformed mapping from simple and elegant cartography into a powerful tool for synthesizing, analyzing, and integrating information. For years, archaeologists have used historical maps, topographic maps, aerial photographs, and satellite imagery to do their work, such as finding the Chacoan roads in New Mexico or mapping the Knife River Indian Villages in North Dakota. Archaeologists embrace GIS because the technology easily displays patterns of information and illustrates geographic relationships between sets of data on maps.

Now out of its infancy, GIS has gained widespread acceptance as a tool for management and analysis. Current GIS technology brings together data and maps in a way not possible a decade earlier, whether the scale of analysis is archaeological survey or excavation. Modeling and statistical analysis are not new to archaeology. But today's archaeology can incorporate many layers of data within a GIS to create models that help predict where sites might be located and how erosion will affect ruins. Archaeologists can resurrect old data sets and apply them to new problems. The outcome might be a statistical model that maps change through time or a map that shows the location and density of artifacts in an ancient village. GIS is but one of the many tools of modern technology that also include standardized electronic topographic maps, geographic imagery, and cultural and environmental data layers. The development of digitized archaeological data and its availability in electronic formats motivated archaeologists to adopt the tools of GIS for management. National Park Service archaeologists can use digital archives, artifact data, and databases to help protect, understand, and explain the value of archaeological sites nationally. GIS map layers might include archaeological data, basemaps, imagery, and different kinds of environmental information. Archaeologists use the maps to plan, research, and monitor sites. In this way, GIS and cultural resource databases assist long-range planning, budgeting, and decision making at all levels.

With government under pressure to show the economic value of its services and resources, GIS has increased productivity, data accuracy, service, and public education. This includes faster response to data requests and to emergencies such as wildfires, vandalism, and looting of archaeological sites. Archaeologists now routinely collect data and transfer it to GIS maps the same day. Direct downloads save time, reduce human error, and improve data interpretation. The future role of GIS in archaeology should enhance resource protection, public education, and stewardship in our national parklands.

Weights of Evidence Analysis for Cultural Resource Site Prediction and Risk Assessment

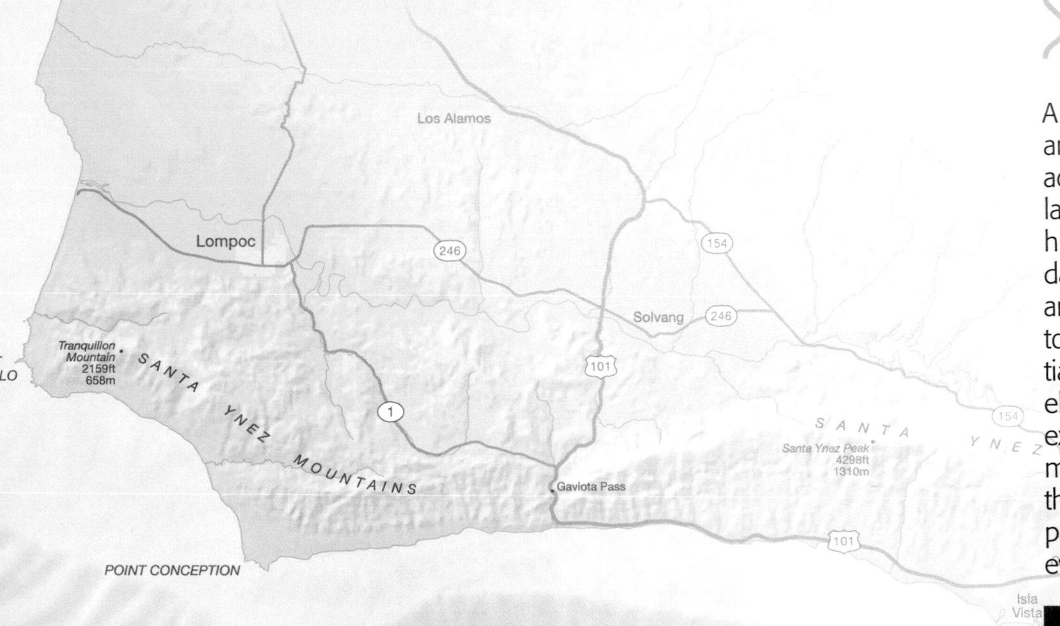

A century of ranching, grazing, and storm and wave erosion is threatening fossils, artifacts, and ancient American Indian villages on Santa Rosa Island near Los Angeles. The fifty-five-thousand-acre island, part of Channel Islands National Park, supported a vibrant culture for most of the last ten thousand years. Archaeologists have documented more than six hundred sites, and human remains found there are among the oldest known in North America. Soil erosion and damage to streams from cattle and nonnative deer and elk all combine to threaten cultural and natural resources. To protect the island, the park service used GIS and other technology to identify and model the areas most at risk. The park service also created a model of potential prehistoric sites. Using GIS software, the park service studied vegetation, geology, and elevation data and reviewed information about roads, streams, pastures, and fire history. And experts in these fields of study offered their views concerning the biggest threats. The two models helped the park service produce two maps. The first illustrated erosion threats and the second showed possible sites of prehistoric settlement on the island. In its findings, the park service concluded that Tobler's First Law of Geography held true: "Everything is related to everything else, but nearer things are more related to each other than distant things."

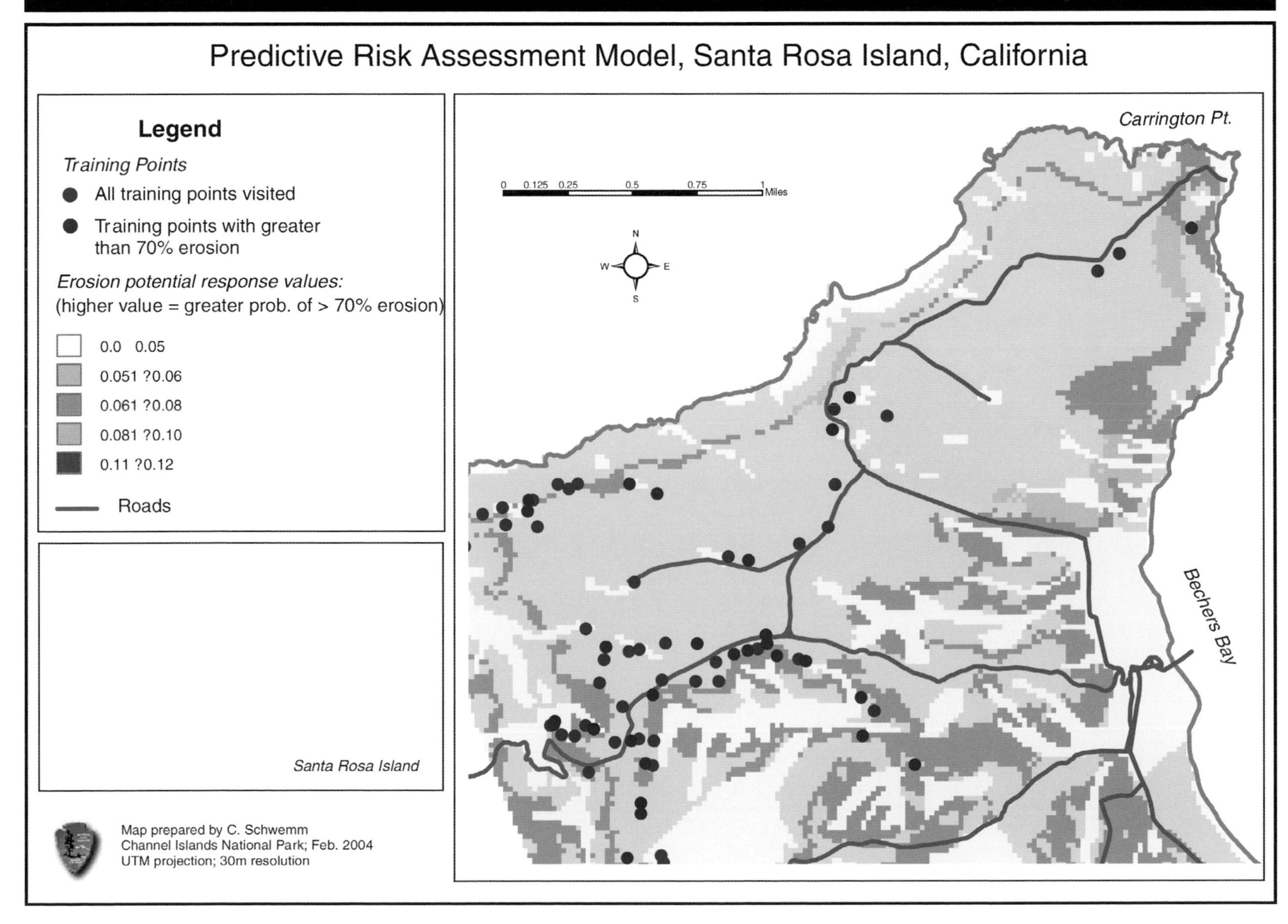

This map displays the potential risk of erosion on a portion of Santa Rosa Island in Channel Islands National Park. Erosion risk is highest in dark green areas.

Clay Tobacco Pipe Stems in New Towne, Jamestown Island

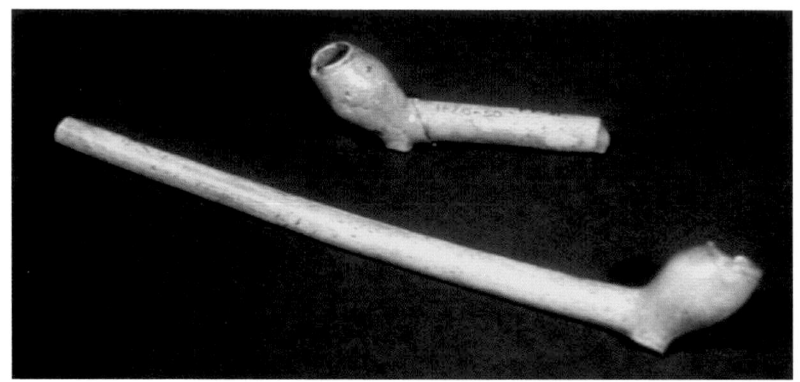

COLONIAL WILLIAMSBURG FOUNDATION

English settlers from the Jamestown colony in Virginia have left us a blueprint of urban growth, if clay tobacco pipe stems found in the "New Towne" area are any indication. Widespread excavations of the first permanent English colony nearly four hundred years ago have turned up an enormous number of artifacts at the New Towne site, pipe stems among them. Updating earlier findings, Colonial National Historical Park imported archaeological site data into a geographic database to show the numbers and sizes of pipe stems found in one-hundred-foot-square grids overlaid on the New Towne area of Jamestown Island. The park service, which administers the site together with the Association for the Preservation of Virginia Antiquities, assigned a range of dates to the pipe stems, based on their bore diameter and on J. C. Herrington's 1954 archaeological work "Dating Stem Fragments of Seventeenth and Eighteenth Century Clay Tobacco Pipes." The findings help plot the haphazard course of New Towne settlement, daily life, and later abandonment in the seventeenth century.

CHARLES D. RAFKIND, NPS PHOTO

Dave Frederick of the National Park Service uses a handheld GPS unit in a marsh on Jamestown Island.

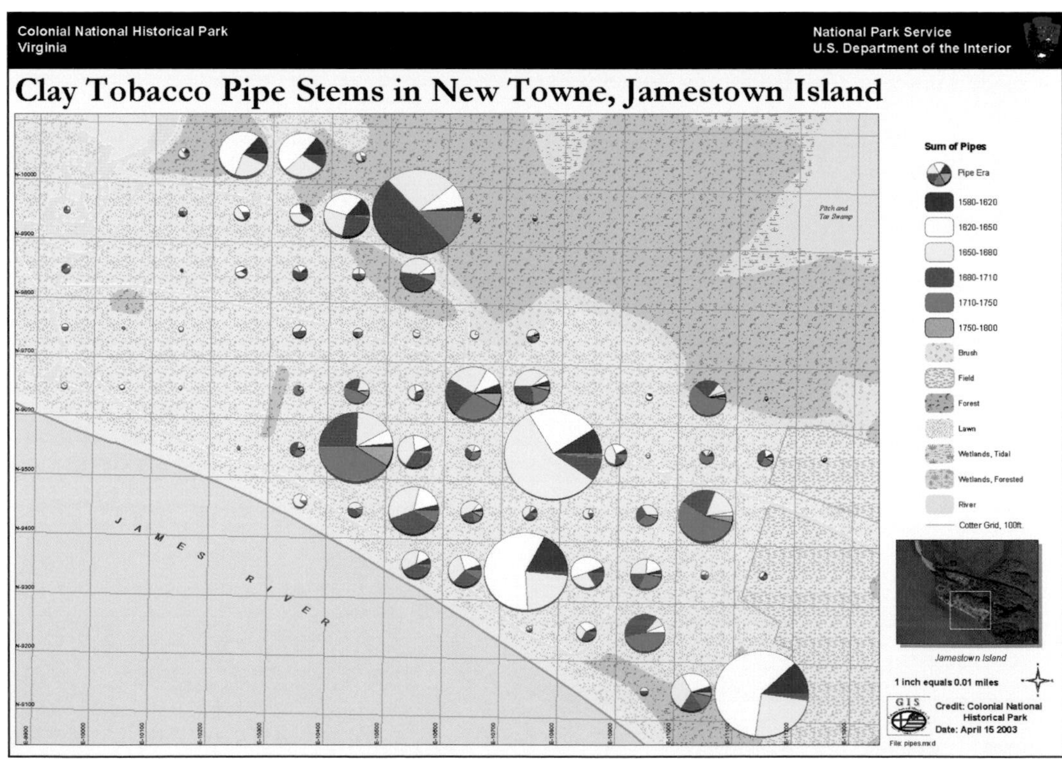

Pie charts display the sum of pipe stems belonging to time periods from the "pipe era" from 1580 to 1800 on Jamestown Island.

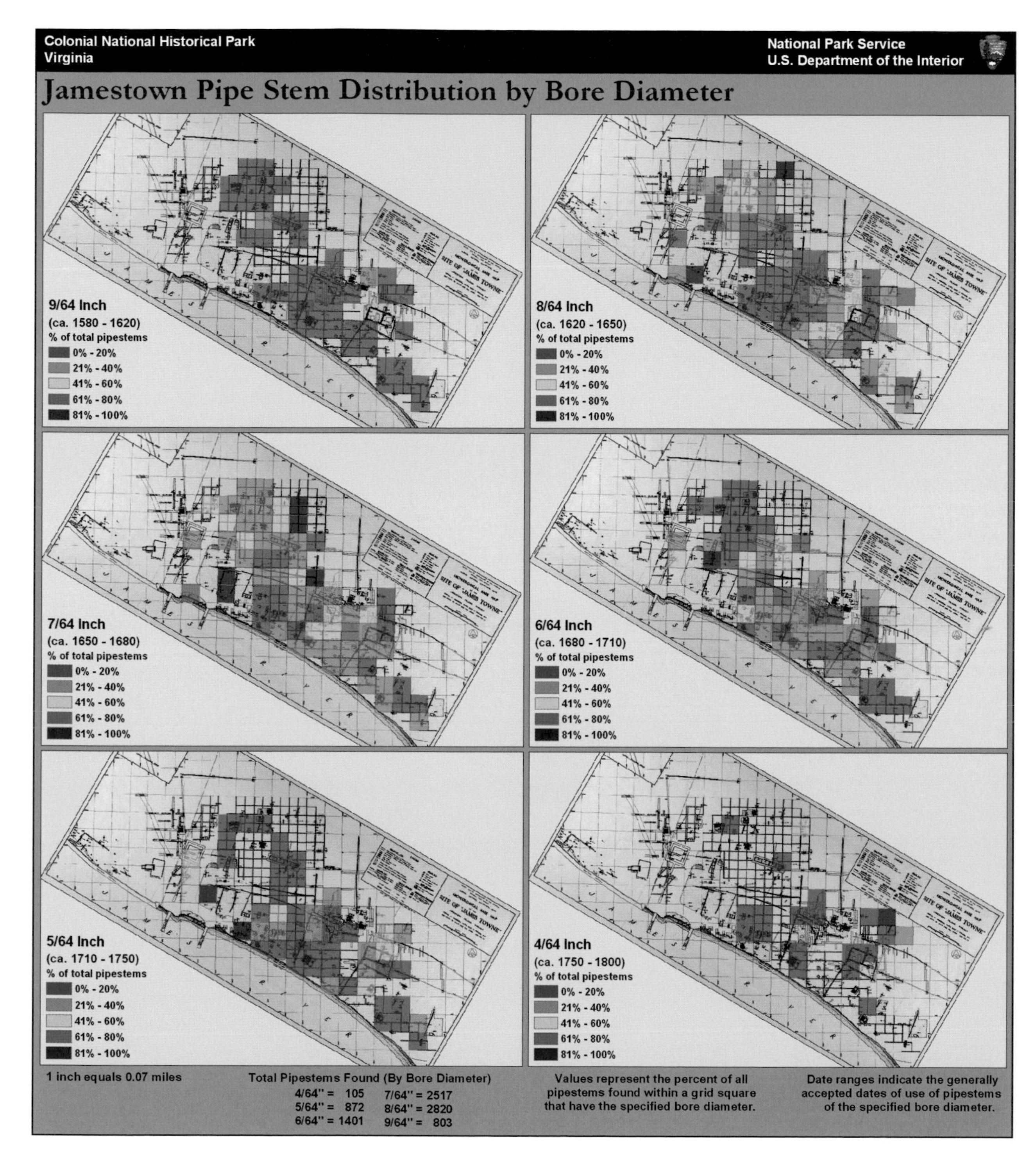

This map displays the percentage of all pipe stems found within a grid square that have a specified bore diameter. The year ranges indicate generally accepted periods during which settlement residents used the specified bore diameter.

CHARLES D. RAFKIND, NPS PHOTO

COLONIAL WILLIAMSBURG FOUNDATION

In historic Jamestown, bricks outline an archaeological site in the scene at the top. Dr. Audrey Horning of the Colonial Williamsburg Foundation works at an excavation, above.

Managing Cultural Resources with GIS: Museum Collections

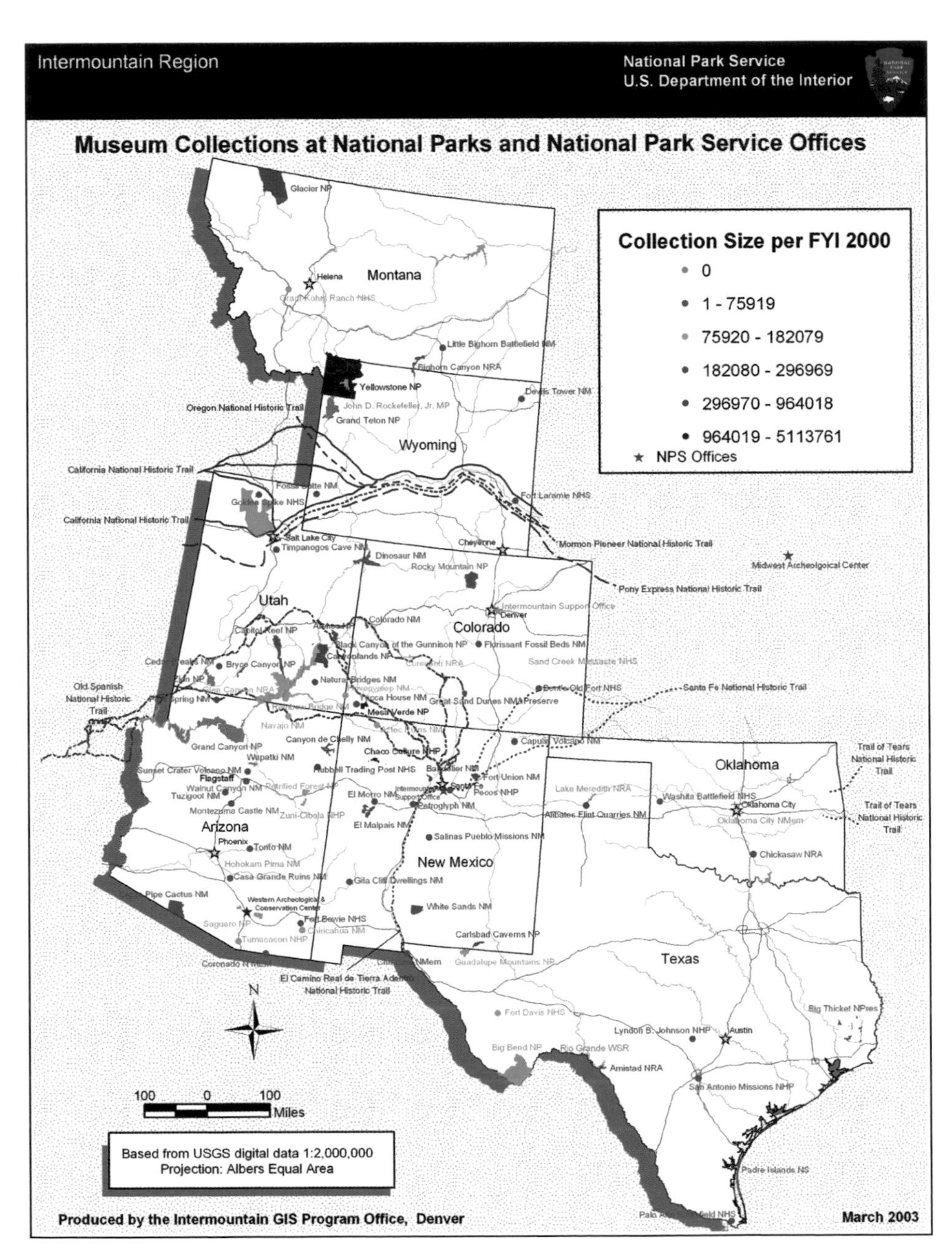

Display of museum collections in the National Park Service Intermountain Region helps the park service plan the care of artifacts.

The preservation and care of artifacts and other museum collections, including proper storage of collections not on exhibit, is a top priority in the National Park Service. To identify the need for new museums, the park service produced a series of GIS maps to document the size and location of its collections originating from the Intermountain Region. The park service knew it had twenty-three million museum items in the region's collections across eight states and on loan throughout the world. But it had never before created a graphic display of the data. Three GIS maps showed that many national and international repositories had collections from thirty-five or more parks in a single location. The data clearly shows active and widespread research using collections loaned from the Intermountain Region. The GIS maps help the park service decide the best ways to manage, lend, store, and improve the care of museum collections for the benefit, education, and enjoyment of people the world over.

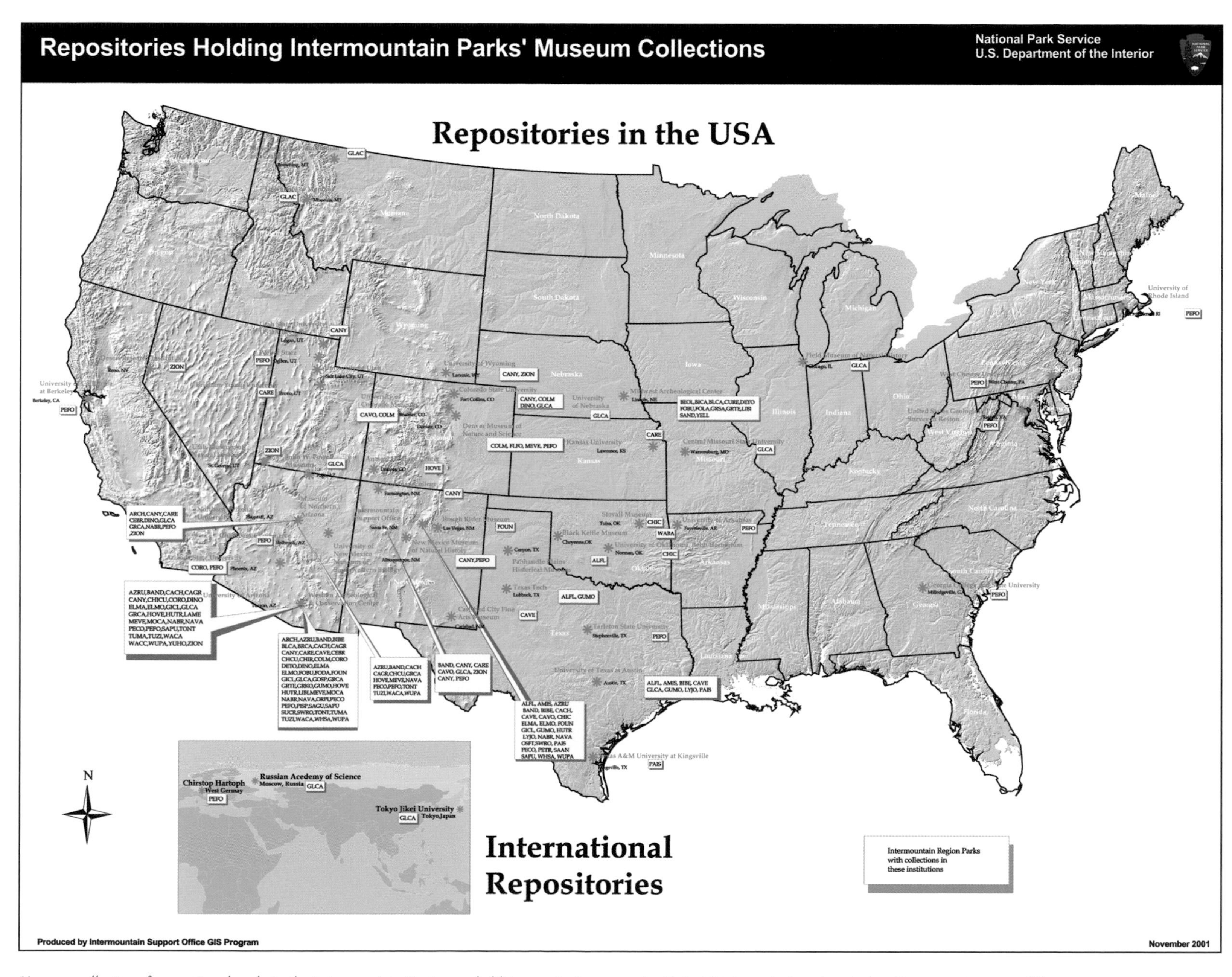

Museum collections from national parks in the Intermountain Region are held in repositories across the United States and abroad, including Germany, Japan, and Russia.

Archaeology and GIS in Western Arctic National Parklands

The National Park Service carried out a cleanup effort in 2003 to remove drums scattered on the north coast of Bering Land Bridge National Preserve.

The four parks that comprise Western Arctic National Parklands in northwest Alaska feature some of the most remote wilderness in the United States. Bering Land Bridge National Preserve, Cape Krusenstern National Monument, Kobuk Valley National Park, and Noatak National Preserve stretch for more than two hundred miles, from the Chukchi Sea coastline to the interior mountains of the western Brooks Range. National park archaeologists heavily depend on GIS to manage more than eleven million acres in these coastal plains, lagoons,

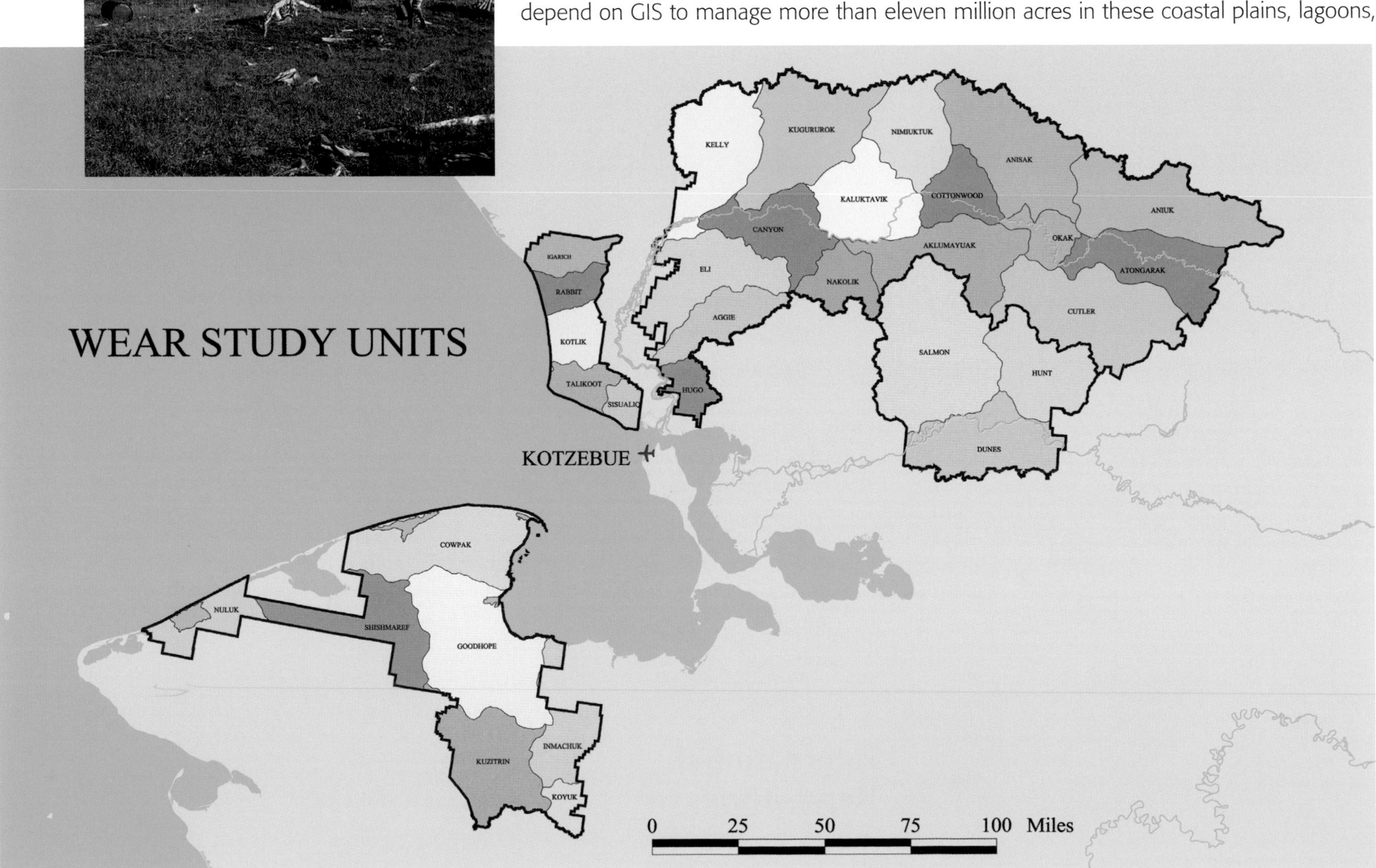

Archaeological study units in the Western Arctic National Parklands of Alaska.

sand dunes, rivers, and mountains. The park service uses GIS to display archaeological sites, cultural uses, travel routes, native lands, and other themes on U.S. Geological Survey maps. Archaeologists use the maps to assess environmental threats, sort new information, and plan for the future. In one case, GIS maps showed the location of drums of hazardous waste threatening archaeological sites. Equally important, the four national parks use GIS to increase the visual impact of posters and slide shows in public and professional settings.

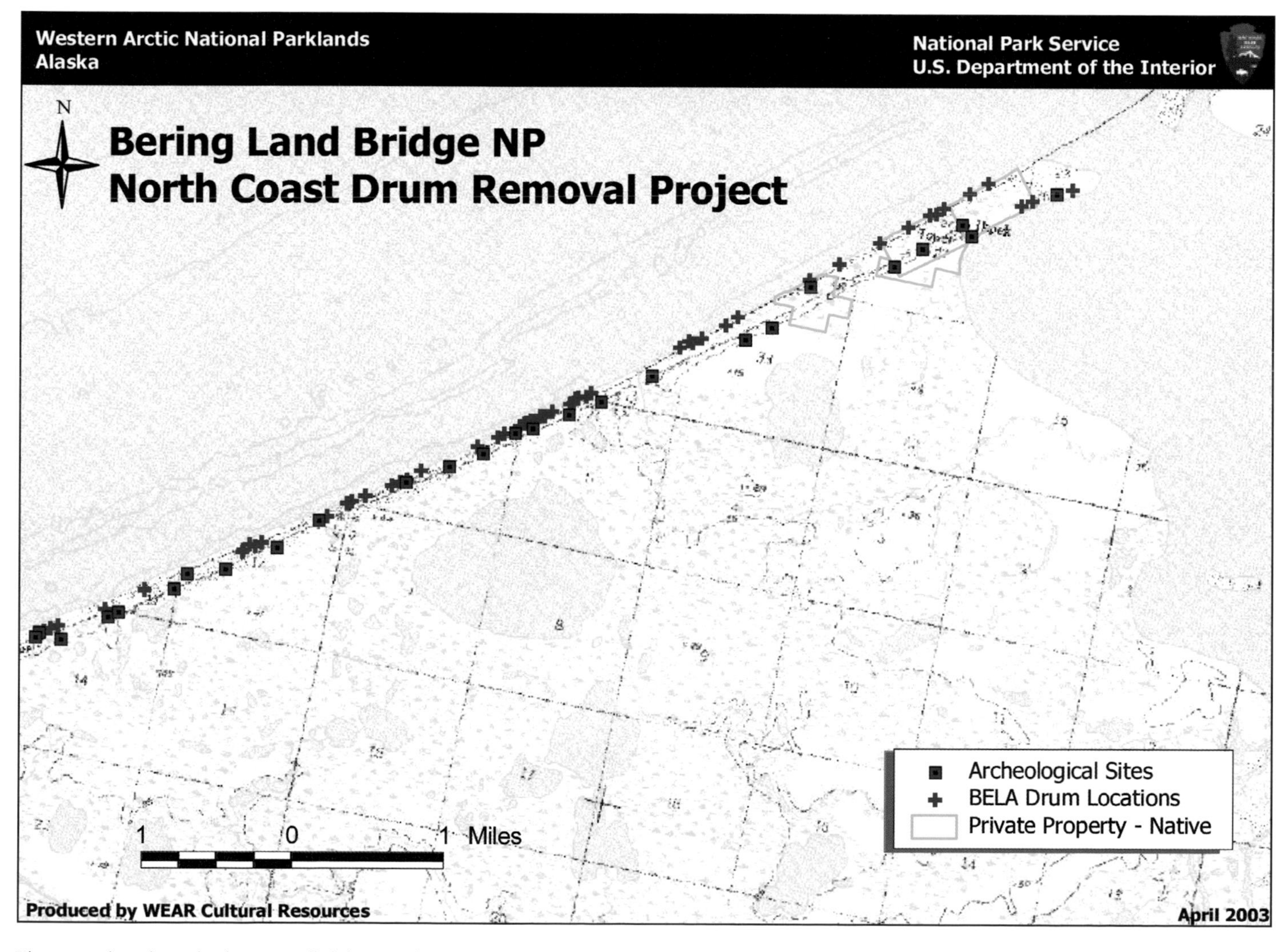

The geographic relationship between spilled drums and archaeological sites on the north coast of Bering Land Bridge National Preserve.

The remains of a historic or late-historic subterranean house excavated on the Chukchi Sea coast of Bering Land Bridge National Preserve.

Chapter 6

Then and Now: Historic Preservation

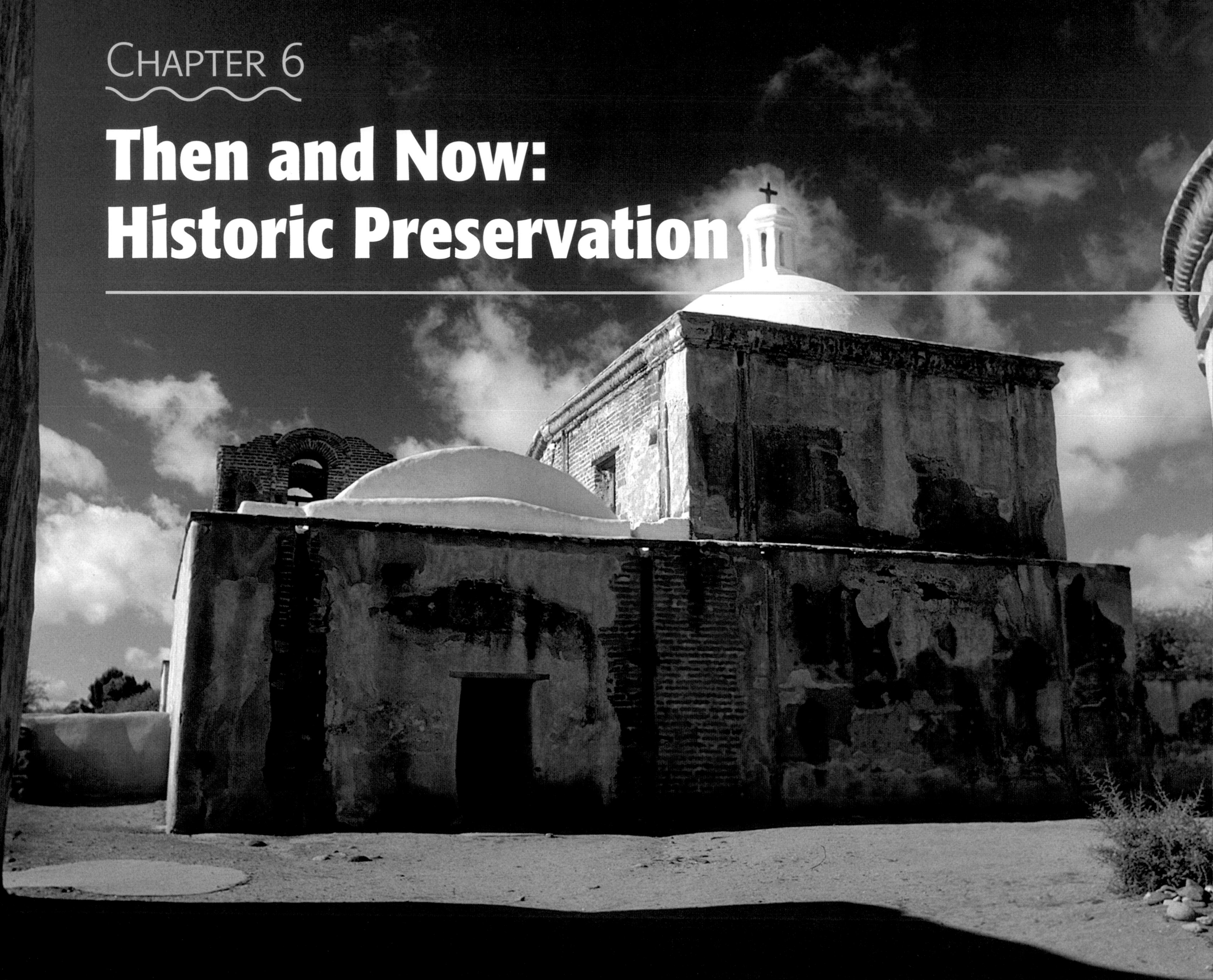

The national park idea has been nurtured by each succeeding generation of Americans. Today, across our land, the national park system represents America at its best. Each park contributes to a deeper understanding of the history of the United States and our way of life; of the natural processes which have given form to our land; and to the enrichment of the environment in which we live.

George B. Hartzog Jr., National Park Service director, 1964–1972

The National Historic Preservation Act became law on October 15, 1966. Congress noted "the historical and cultural foundations of the Nation should be preserved as a living part of our community life and development in order to give a sense of orientation to the American people." Preserving archaeological sites, historic buildings, and landscapes must take place within the context of the evolution of American communities. Making this concept a reality is no easy task. It involves balancing community needs for new housing, sewage treatment plants, hospitals, and shopping centers with the need to protect historic courthouses, ancient ruins, and Civil War battlefields. To succeed, we try to understand the relationships between community values and historic resources. We often define these relationships by the distribution of geographic features. GIS is the perfect tool to highlight these spatial relationships.

National parks are, in a sense, communities. They require roads, visitor centers, pipelines, bridges, and utilities, just like the hometowns we leave behind for vacations in our national parklands. Like many communities, parks are coping with the pressures of increased population as visitors bring more RVs, boats, tents, stoves, and pets into the parks. The National Park Service preserves parklands for our benefit and enjoyment, as part of its mission. Yet the popularity of national parks has strained the bureau's ability to protect cultural resources. To ease this stress, park managers use high-tech tools, including GIS, to balance visitor use and cultural resource preservation. For example, GIS software can create maps that show the spatial relationship between a proposed campground and a historic site, or between a planned highway and a Revolutionary War battlefield.

With each success, the park service increasingly embraces GIS to define relationships between park management needs and historic preservation. At Fort Smith National Historic Site in Arkansas, researchers combined GIS with historic maps to identify archaeological sites. At Salinas Pueblo Missions National Monument in New Mexico, historians used GIS to help stabilize Pueblo and Spanish ruins. Wilson's Creek National Historic Battlefield mapped the possible location of two artillery positions to interpret the Civil War battle for visitor education and enjoyment. In the Northeast, the park service linked cultural resource databases with GIS to deliver a complete body of information on each cultural resource. And the bureau's Cultural Resources GIS Facility used GIS technology to survey more than eight hundred sites related to the Revolutionary War and War of 1812, to assess their condition. In these and other projects nationwide, GIS has made important links between cultural resources and park management. These relationships are essential if park managers are to make historic preservation a living part of the communities they serve.

GIS in Ruins and Historic Structures Preservation at Salinas Pueblo Missions National Monument

A close-up of the Abo Mission as seen on a georectified aerial photograph. The large-scale orthophoto mosaic helps locate and map various features and sites at the mission.

Austere yet beautiful reminders of early contact between Pueblo Indians and Spanish explorers are visible in the ruins at Salinas Pueblo Missions National Monument in New Mexico. Established in 1980 through the combination of two state monuments and a national monument, the present monument boundaries encompass eleven hundred acres. The decades-long work to preserve the rich history of the Salinas Valley now includes GIS mapping of prehistoric and historic features to manage and automate these resources. From stored databases, archaeologists can retrieve geographic information and field assessment reports and select from thousands of scanned photographs as they work to preserve the ruins of four missions and surrounding pueblos. GIS projects illustrated in these maps involve digitized, scanned drawings of pueblo and mission structures, GPS data collection, and conversion of computer-aided-design data sets. GIS mapping, databases, and large-scale aerial photography assist archaeologists and historians as they learn more about the history of the Pueblo Indian trading communities that thrived in this remote frontier country until the late seventeenth century, when drought, famine, cultural conflict, and warfare drove the Spanish and their Indian allies south to El Paso.

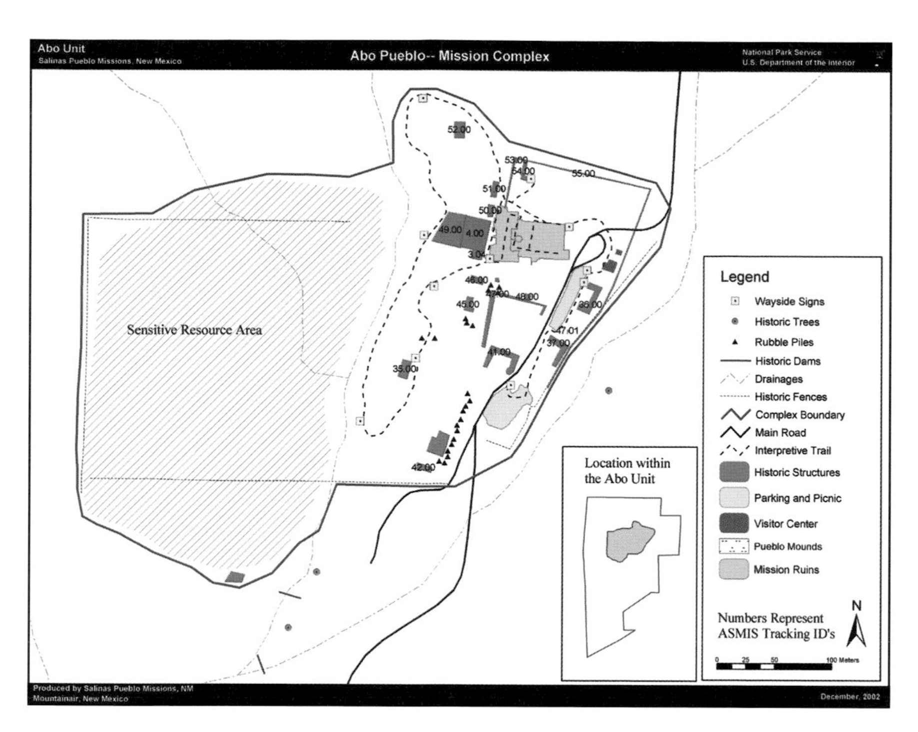

A GIS map of the Abo unit of the national monument illustrates the geographic relationship of prehistoric, historic, and modern features, including dams, trees, roads, trails, and wayside signs.

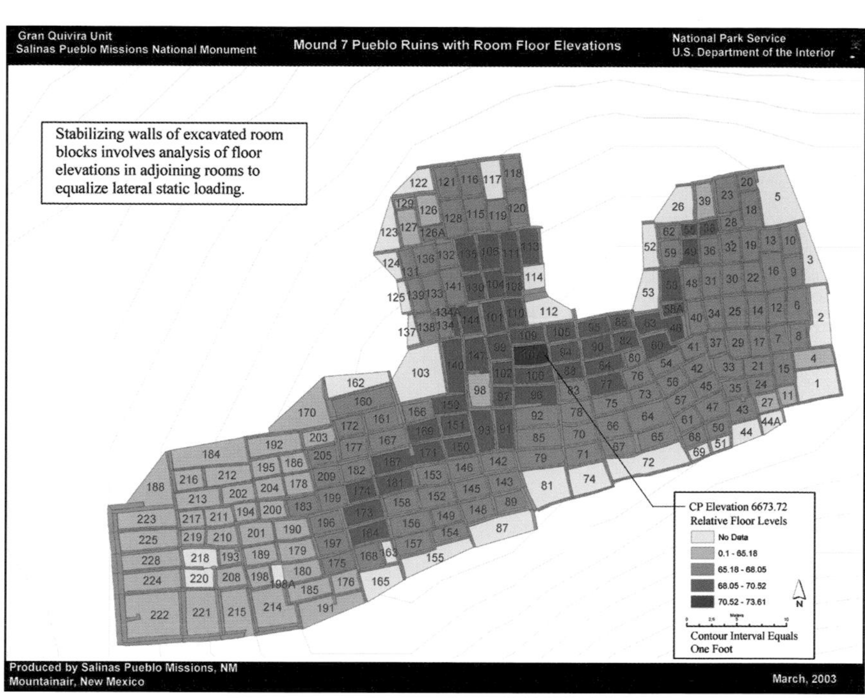

This map illustrates part of the project to preserve the excavated ruin of Mound 7 at the Grand Quivira unit. Earlier excavations had left fragile walls exposed to the elements, and backfilling is being considered as a way to help stabilize the walls and preserve the ruin.

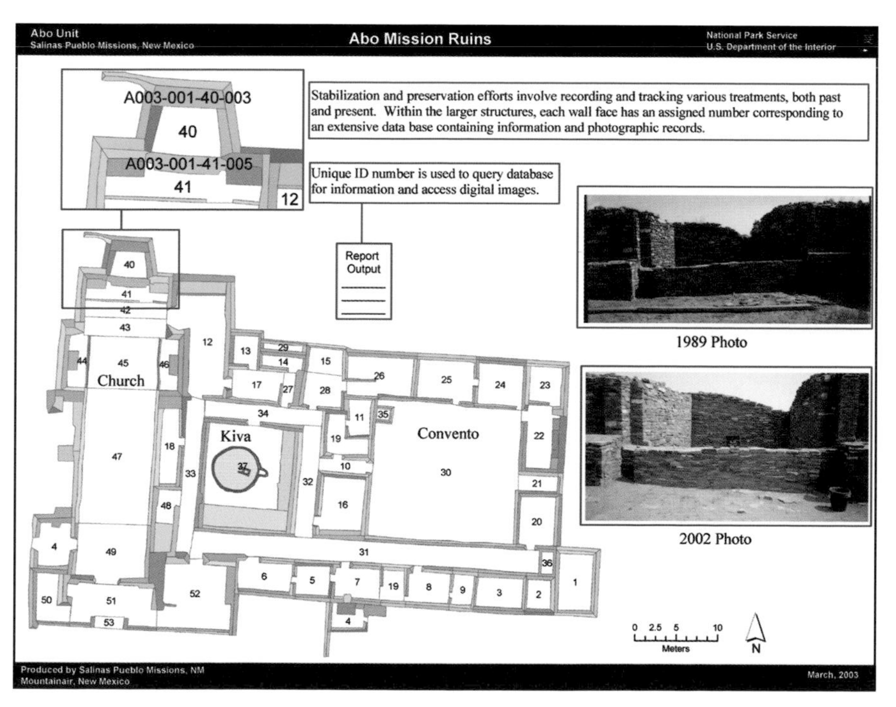

The assigned numbers of individual wall faces on this map of the Abo Mission ruins are used to join digital maps with attribute data (information and photographic records) contained in a database.

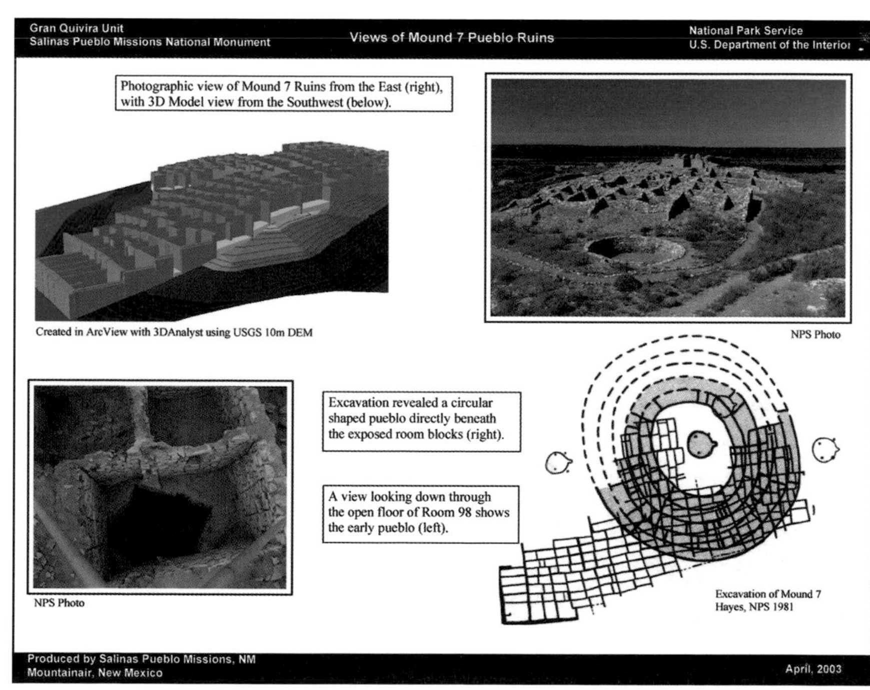

This map contains four different types of views of Mound 7 at the Grand Quivira unit, using photographs, a three-dimensional model, and a sketch.

Using Historic Maps to Anticipate and Manage Archaeological Resources in Fort Smith National Historic Site, Arkansas

Visitors can still see the remains of the first Fort Smith, shown here.

Fort Smith National Historic Site in Arkansas embraces the remains of two frontier forts and a federal courthouse. The government opened the first Fort Smith in 1817 to keep the peace between local Osage and Cherokee who had been forced from their ancestral lands in the Southeast. The second fort opened in 1838 to deal with ongoing westward migration of American Indians and settlers. The Federal Court for the Western District of Arkansas replaced the fort in 1871, after the Civil War. The court served as a buffer between outlaws and peaceful citizens and handled legal matters in the Indian Territory and western Arkansas. The site today reminds visitors of eighty turbulent years in the history of federal policy toward American Indians. The National Park Service has completed a mapping program using GIS technology that included evaluation of more than seventy-five historic maps. The project resulted in a digital atlas of more than thirty-five themes, each showing a different kind of geographic information. The atlas will help park managers and researchers avoid damage to archaeological features in the park. Displays of the mapping process will offer visitors a new and easier way to understand park history and the role of the forts in the development of the West.

Areas of archaeological investigation are shown in relation to features of the first Fort Smith from 1817 to 1824.

GIS map of Fort Smith National Historic Site, displaying the relationship between historic features, possible archaeological remains, and the West Fort Smith Settlement from 1906 to 1957.

Archaeological areas are shown according to their sensitivity, from low to high.

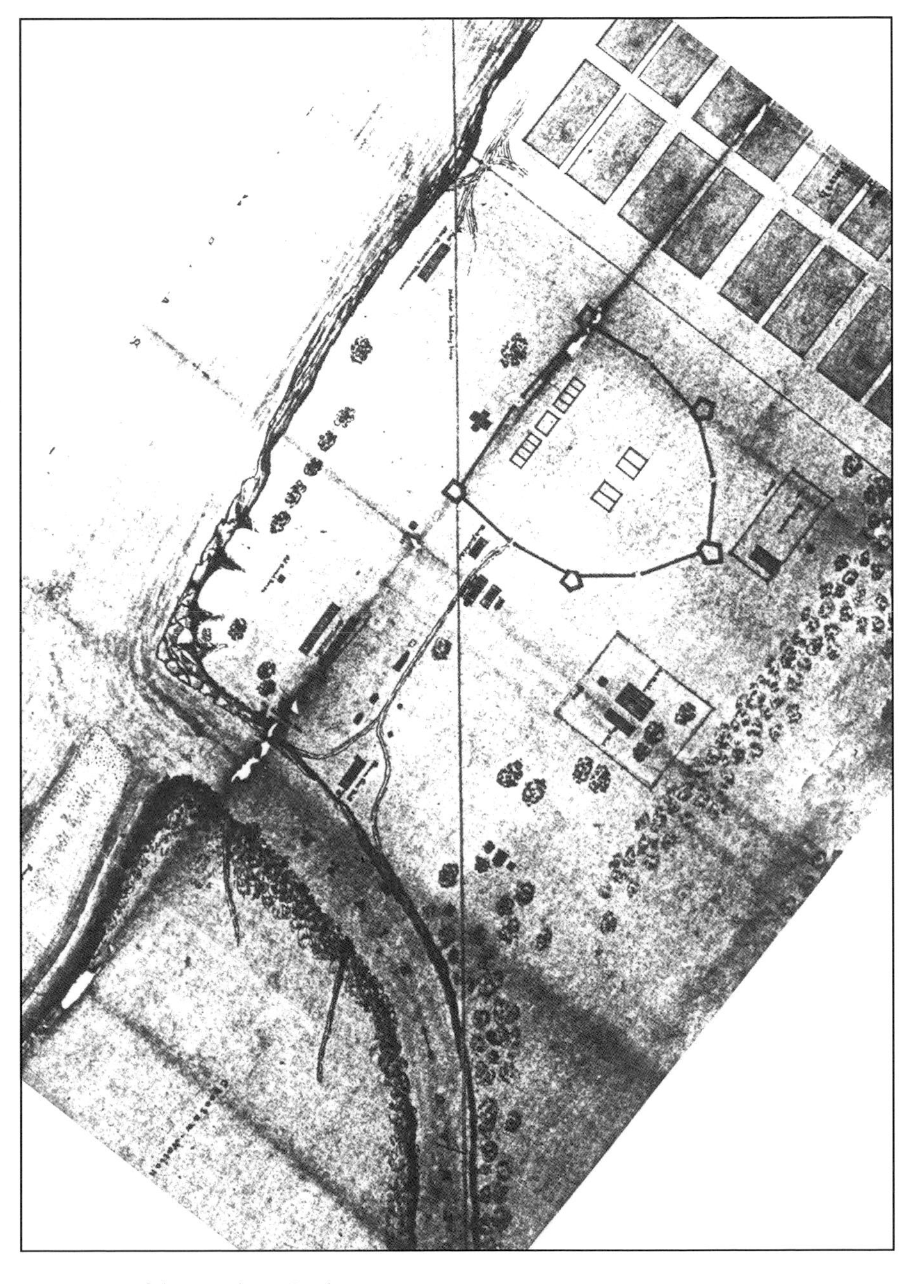

Historic map of the second Fort Smith.

Using GIS to Identify Possible Artillery Positions and Manage Artifact Data at Wilson's Creek National Battlefield

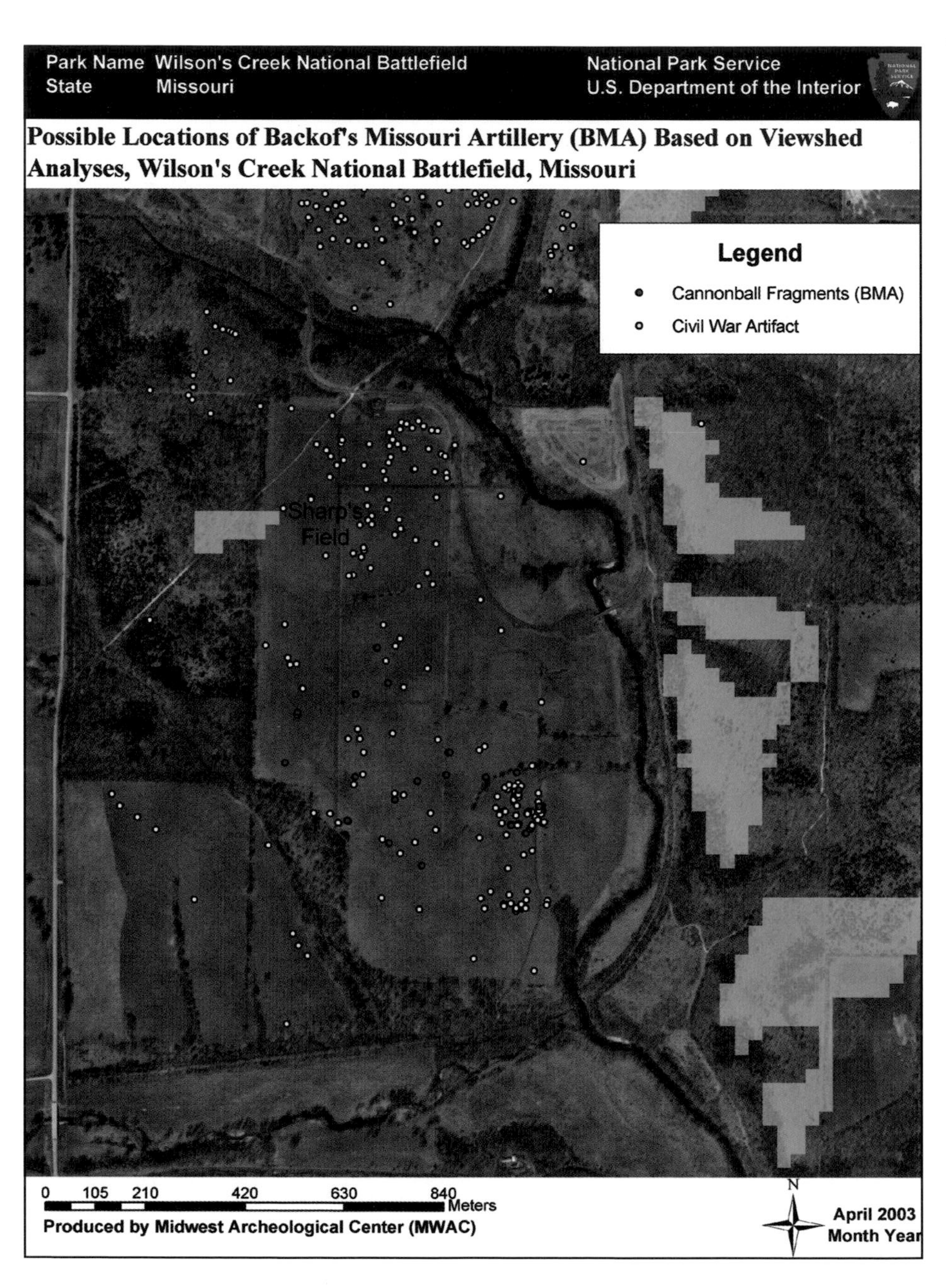

The areas in green identify possible Union artillery positions at Wilson's Creek National Battlefield in Missouri, based on a viewshed analysis. The map also displays the distribution of cannonball fragments and other Civil War artifacts.

The first major battle of the Civil War west of the Mississippi River erupted August 10, 1861, along a quiet brook near Springfield, Missouri. There at Wilson's Creek, an intense firefight raged for hours between Union and Confederate troops. More than 140 years later, archaeologists from the Midwest Archeological Center are using GIS to more accurately pinpoint artillery positions and record the locations of bullets, shell fragments, and personal belongings that show where opposing lines fought, and men died. GIS mapping helped historians reconstruct events at Wilson's Creek National Battlefield Park. In one instance, a GIS analysis of shell fragments suggested a more accurate position for Union artillery that surprised Confederate cavalrymen at the south end of Sharp's Field at dawn. In another, GIS mapping showed where a Confederate battery routed a Union brigade at the north end of the field. This sort of analysis has become increasingly necessary as a way to clarify and correct the historical record. GIS mapping has helped the park service better understand and interpret park history and present a more accurate and sophisticated view of the battle to the public.

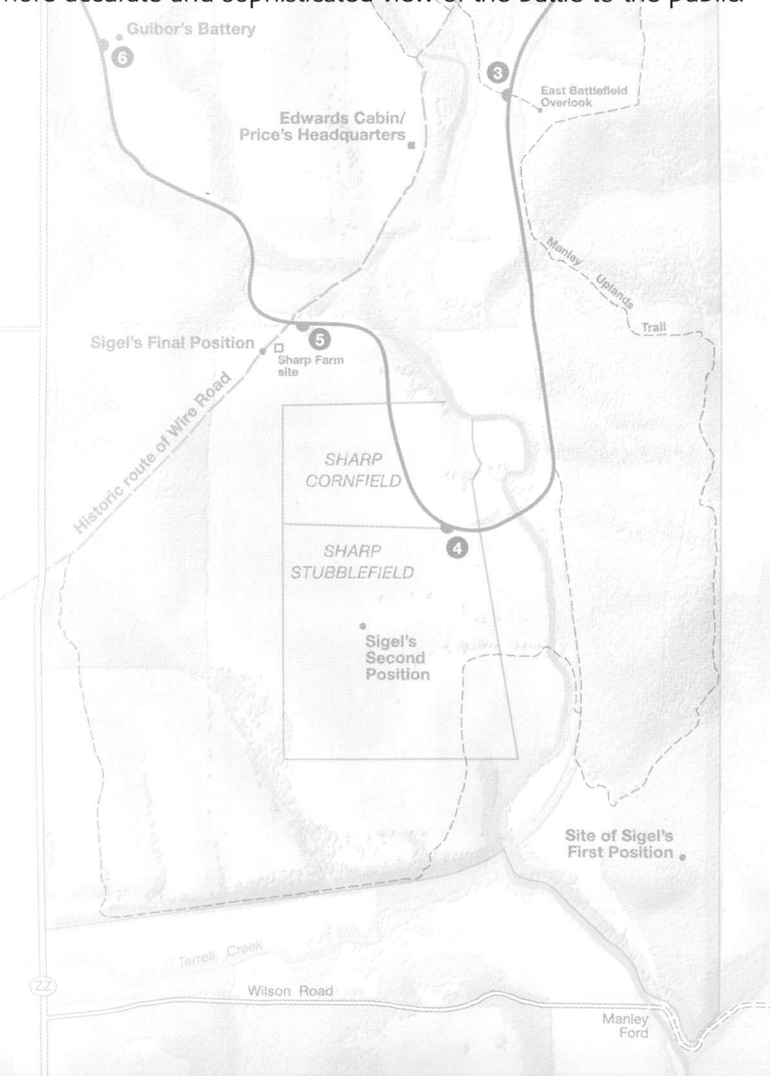

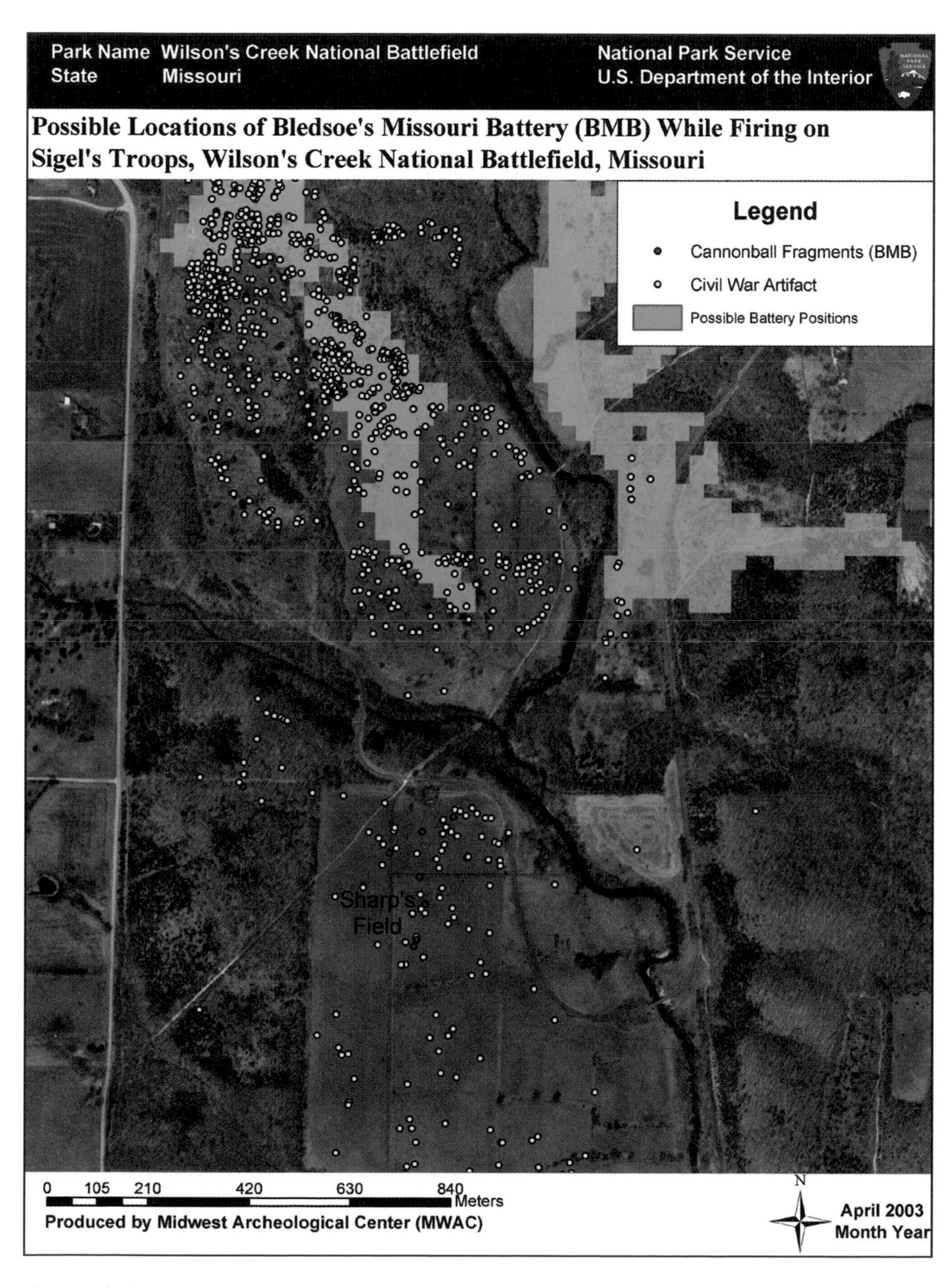

This viewshed analysis illustrates the positions from which Confederate Captain Hiram M. Bledsoe's Missouri battery could have fired on Union Colonel Franz Sigel's brigade in Sharp's Cornfield.

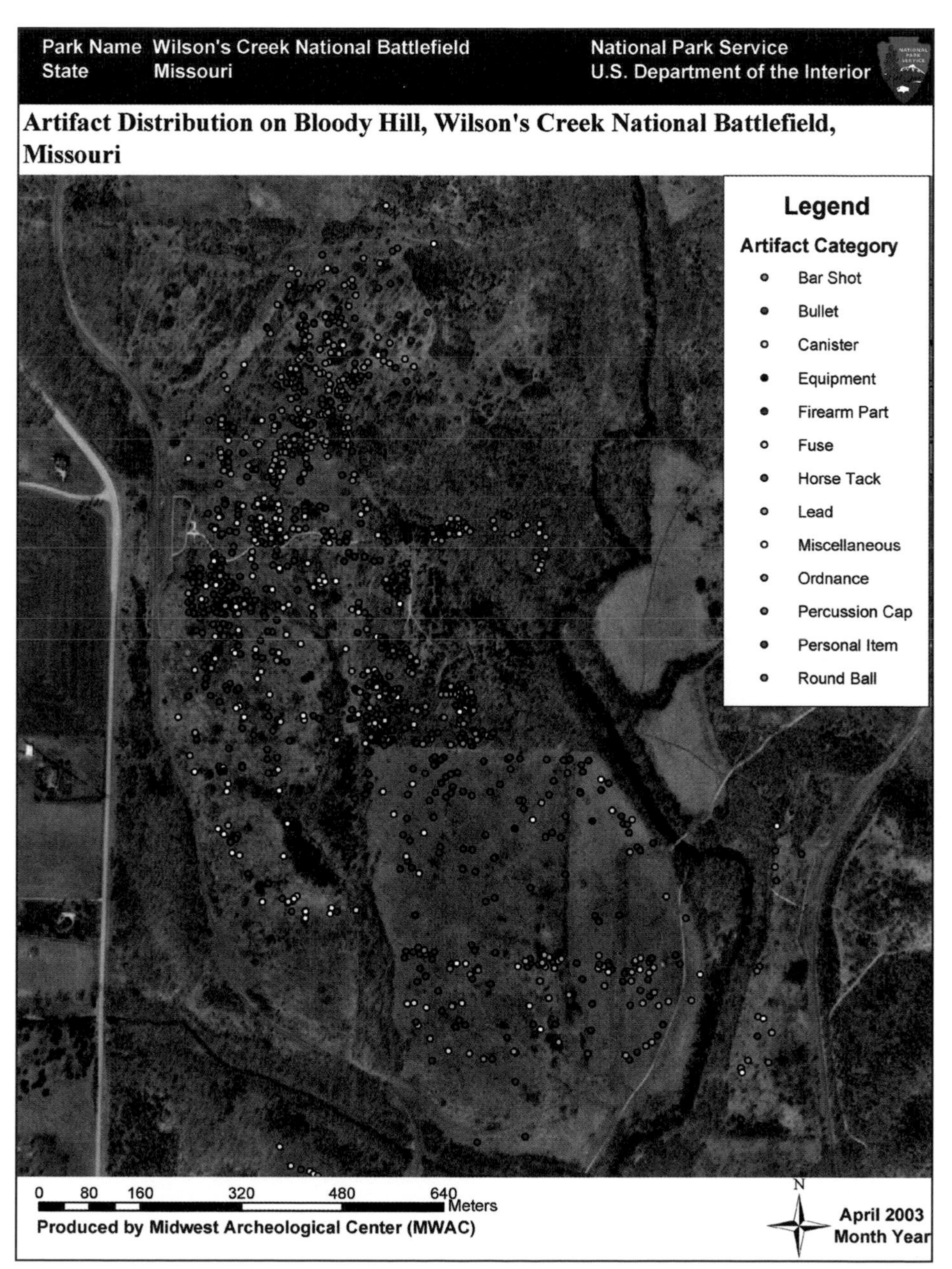

The distribution of musket balls, equipment, gun parts, minie balls, horse tack, and other artifacts on Bloody Hill.

Linking Cultural Resource Databases through GIS

GIS technology integrates databases of cultural resources at Hopewell Furnace National Historic Site in Pennsylvania, above, and at Appomattox Court House National Historical Park in Virginia, on facing page.

The National Park Service relies on databases full of cultural information about archaeology, landscapes, historic buildings, museum collections, and characteristics and customs of different peoples. But until now, the park service had no easy way to retrieve the data from a single location. Now, thanks to GIS, the park service can link the information kept in separate databases, putting data at the fingertips of park planners and managers. North Carolina State University's Center for Earth Observation developed a method to determine the geographic relationship between each feature in the databases, using an x,y coordinate system. The center then built a geographic information system by linking information from each database to the corresponding features. The GIS has made the information readily available, allowing research queries and analyses based on information drawn from any of the databases.

Graduate students at the Center for Earth Observation at North Carolina State University analyze and model spatial dynamics of the hemlock woolly adelgid, a forest pest found in the southern Appalachians.

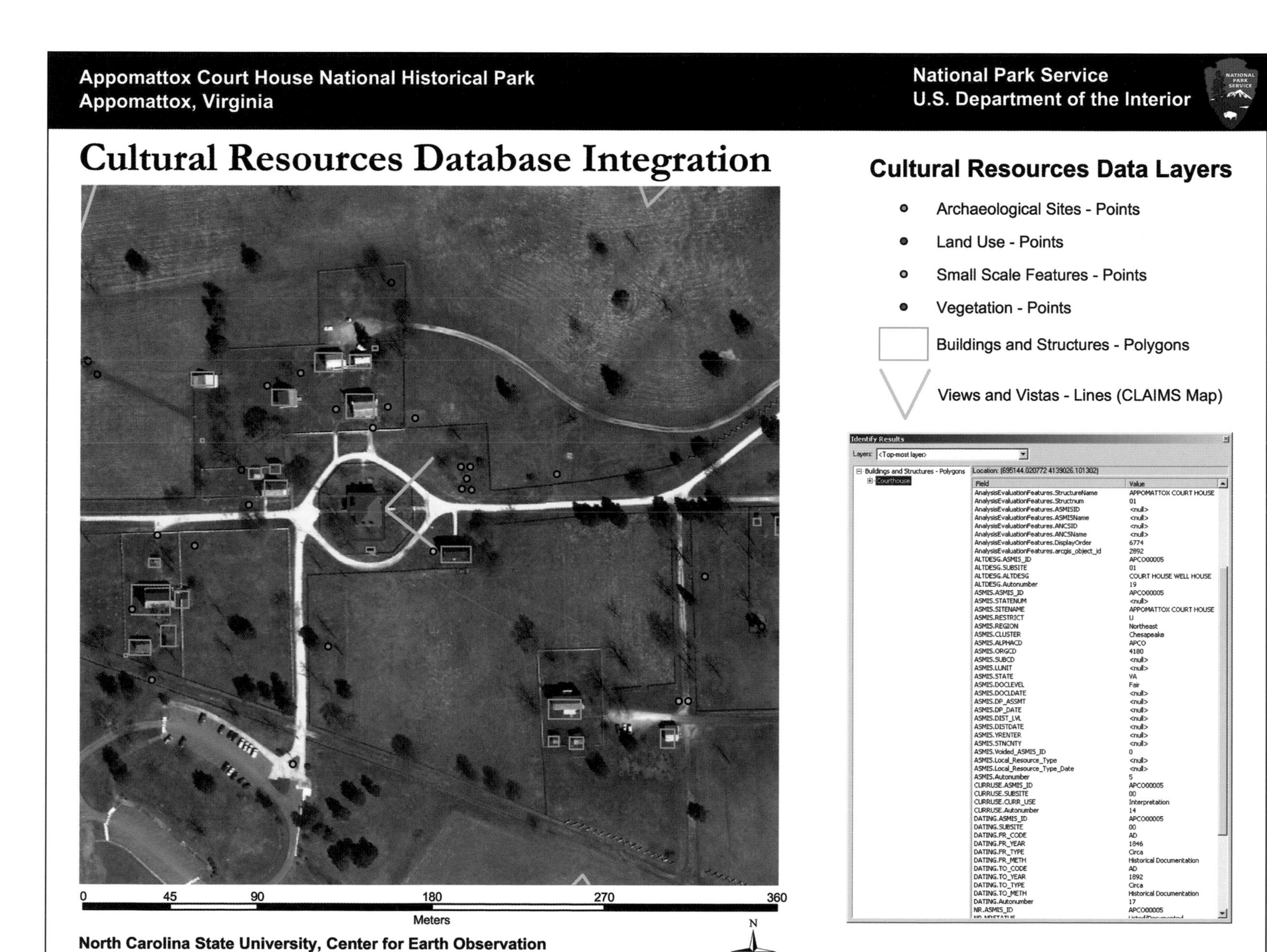
Appomattox Court House National Historical Park
Appomattox, Virginia
National Park Service
U.S. Department of the Interior
Cultural Resources Database Integration
Cultural Resources Data Layers
Archaeological Sites - Points
Land Use - Points
Small Scale Features - Points
Vegetation - Points
Buildings and Structures - Polygons
Views and Vistas - Lines (CLAIMS Map)
Identify Results
Layers: <Top-most layer>
Buildings and Structures - Polygons
Courthouse
Location: (695144.020772 4139026.101302)
Field Value
AnalysisEvaluationFeatures.StructureName APPOMATTOX COURT HOUSE
AnalysisEvaluationFeatures.Structnum 01
AnalysisEvaluationFeatures.ASMISID <null>
AnalysisEvaluationFeatures.ASMISName <null>
AnalysisEvaluationFeatures.ANCSID <null>
AnalysisEvaluationFeatures.ANCSName <null>
AnalysisEvaluationFeatures.DisplayOrder 6774
AnalysisEvaluationFeatures.arcgis_object_id 2892
ALTDESG.ASMIS_ID APCO00005
ALTDESG.SUBSITE 01
ALTDESG.ALTDESG COURT HOUSE WELL HOUSE
ALTDESG.Autonumber 19
ASMIS.ASMIS_ID APCO00005
ASMIS.STATENUM <null>
ASMIS.SITENAME APPOMATTOX COURT HOUSE
ASMIS.RESTRICT U
ASMIS.REGION Northeast
ASMIS.CLUSTER Chesapeake
ASMIS.ALPHACD APCO
ASMIS.ORGCD 4180
ASMIS.SUBCD <null>
ASMIS.LUNIT <null>
ASMIS.STATE VA
ASMIS.DOCLEVEL Fair
ASMIS.DOCLDATE <null>
ASMIS.DP_ASSMT <null>
ASMIS.DP_DATE <null>
ASMIS.DIST_LVL <null>
ASMIS.DISTDATE <null>
ASMIS.YRENTER <null>
ASMIS.STNCNTY <null>
ASMIS.Voided_ASMIS_ID 0
ASMIS.Local_Resource_Type <null>
ASMIS.Local_Resource_Type_Date <null>
ASMIS.Autonumber 5
CURRUSE.ASMIS_ID APCO00005
CURRUSE.SUBSITE 00
CURRUSE.CURR_USE Interpretation
CURRUSE.Autonumber 14
DATING.ASMIS_ID APCO00005
DATING.SUBSITE 00
DATING.FR_CODE AD
DATING.FR_YEAR 1846
DATING.FR_TYPE Circa
DATING.FR_METH Historical Documentation
DATING.TO_CODE AD
DATING.TO_YEAR 1892
DATING.TO_TYPE Circa
DATING.TO_METH Historical Documentation
DATING.Autonumber 17
NR.ASMIS_ID APCO00005
0 45 90 180 270 360
Meters
N
North Carolina State University, Center for Earth Observation
National Park Service, Northeast Region
April 2003

Mapping the Revolutionary War and War of 1812

Handheld GPS units help fix the location of artifacts, troop positions, fence lines, and other characteristics used to identify and preserve historic battlefields.

Yorktown, Bunker Hill, and New Orleans are getting the once-over as the National Park Service uses high-tech tools to look at the battlefields of the Revolutionary War and War of 1812. With congressional support, the park service set out to identify and preserve significant battlefields for future generations. With help from GIS and related technologies such as global positioning systems, the park service created an online database for scholars, researchers, and the public to ask questions and make suggestions and corrections concerning potential battlefields. After reviewing all the information, experts selected 884 sites in thirty-two states for field surveys. Over the next two years, surveyors from federal and state agencies, universities, museums, and private institutions mapped the battlefields, integrating GIS, the Internet, digital databases, topographic maps, laptops, and GPS. The resulting statistics and maps will shape park service recommendations on battlefield preservation.

Revolutionary War reenactors at Saratoga National Historic Park in New York.

The Revolutionary War & War of 1812 Historic Preservation Study

A study overview and the role of GIS, GPS, database, and internet technologies in generating a report to the U.S. Congress

Cultural Resources GIS Facility, National Park Service, Washington, D.C.

Public Law 104-333, Sec. 603. Revolutionary War and War of 1812 Historic Preservation Study

The Congress finds that -
1) Revolutionary War sites and War of 1812 sites provide a means for Americans to understand and interpret the periods in American history during which the Revolutionary War and Ware of 1812 were fought;
2) the historical integrity of many Revolutionary War and War of 1812 sites is at risk because many of the sites are located in regions that are undergoing rapid urban or suburban development; and
3) it is important, for the benefit of the United States, to obtain current information on the significance of, threats to the integrity of, and alternatives of the preservation and interpretation of Revolutionary War sites and War of 1812 sites.

Transmittal to Congress - Not later than 2 years after the date on which funds are made available to carry out the study, the Director of the National Park Service shall transmit a report describing the results of the study to the Committee on Resources of the House of Representatives and the Committee on Energy and Natural Resources of the Senate.

The Study

Authorized in 1996
Funded in FY 2000 & FY 2001
Modeled after Civil War Sites Study Act of 1990

Follow me!

Study Goals:
* Identify sites
* Rank historic significance of sites
* Assess short- and long-term threats to the integrity of the sites
* Analyze survey data
* Develop preservation alternatives
* Provide broad recommendations for site preservation and interpretation

USS Constitution

Scope of Study

Time Periods Identified
Revolutionary War: April 4, 1775 - September 3, 1783
War of 1812: November 7, 1811 - June 30, 1815

Events Related to Conduct of Wars Only
No monuments, veterans cemeteries, museum objects or places associated with famous persons (e.g. George Washington slept here)

Two Categories of Cultural Resources
* Battle related: e.g. battlefields, underwater archaeological sites of naval engagements
* Associated Historic Properties: e.g. houses, mills, forts, shipyards

Horseshoe Bend National Military Park, Alabama

Old Tenant Meeting House, New Jersey

Identification of Sites

The American Battlefield Protection Program (ABPP) identified states and counties where actions took place.

ABPP and Cultural Resources GIS (CRGIS) staff searched The National Register of Historic Places inventory and identified those properties and sites related to either war.

Details of properties were entered into the Study database.

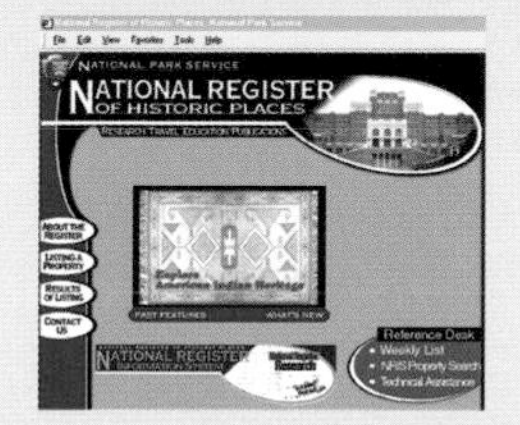

www2.cr.nps.gov/abpp/revwar.htm

To facilitate a period for public and professional comments, a web-based database application was developed using ColdFusion technology allowing individuals to access the Study database and make comments, corrections, and suggest additions to the site list.

As a result of this process, 2,748 sites of known battle actions and historic places associated with either war were identified.

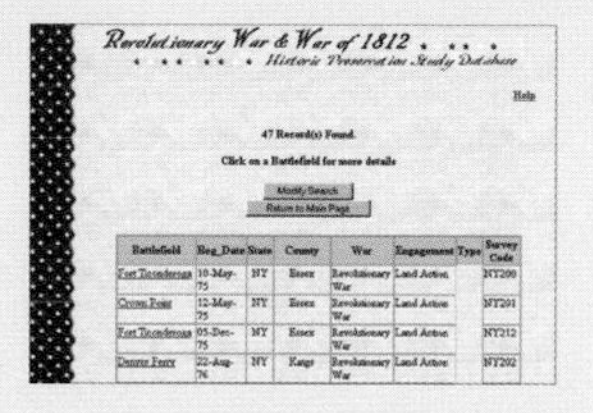

CRGIS Digitizer Application

A highlight of the technology components of this study was a custom-built GIS digitizing application created by John Buckler of CRGIS. The digitizer was created using MapObjects 2.0 and was tailored to meet the digitizing needs of novice GIS users.

The surveyors used the digitizer to create digital boundaries showing the sites' core battle areas and associated study areas.

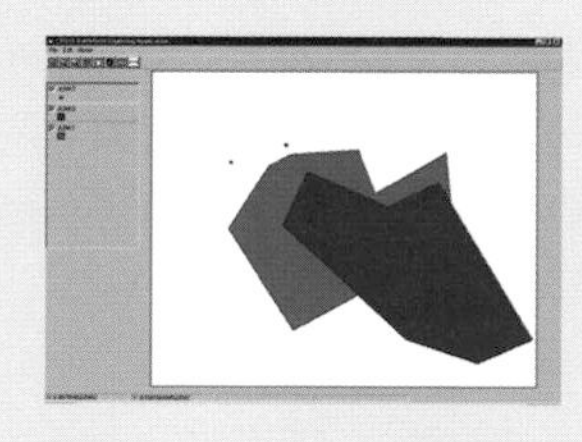

GPS and the Field Survey

CRGIS developed a special data dictionary for the Trimble GeoExplorer 3 which surveyors used in the field. With this data dictionary, surveyors were not only able to record the locations of battlefields and Associated Historic Properties within +/- 5 meter accuracy, but a variety of other relevant landscape and military features and defining attributes such as condition of resource, battlefield name, property type, and land use.

Field Schools

More than 70 individuals were trained in techniques critical to the successful survey of the identified sites. Surveyors were instructed in proper research procedures, how to 'read' a battlefield using military landscape analysis, how to plot thematic features on a map, and informed of the guidelines and requirements of the study.

The main focus of the field schools was the use of various technologies to produce more accurate survey data resulting in detailed information related to the study sites.

The technologies introduced included:

* Global Positioning Systems
* Geographic Information Systems
* Trimble Pathfinder Office 2.7 software
* Custom made GIS application to digitize spatial data
* Custom designed survey database application to record findings

The Advisory Committee

The Revolutionary War and War of 1812 Historic Preservation Advisory Committee, a subcommittee of the National Park System Advisory Board, was made up of scholars and others with a particular interest in the Revolutionary War and the War of 1812. In December 2000, the Advisory Committee completed its evaluation of the relative significance of the sites to the history of the Revolutionary War and the War of 1812.

The Committee's role included the following:

* Identify campaigns and themes of both wars
* Define the criteria for classes of sites (A-D)
* Individually rank battles and Associated Historic Properties
* Review sites for final rankings

From the initial 2,748 sites, 811 were selected as representing the principal historic events of both wars.

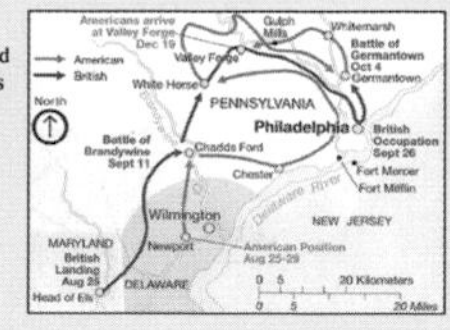

Digital Survey Form

Surveyors used a custom-designed database application to record information related to their research and findings. Information included research and reference materials, contact information, landuse, property characteristics, threats to site, and local planning information.

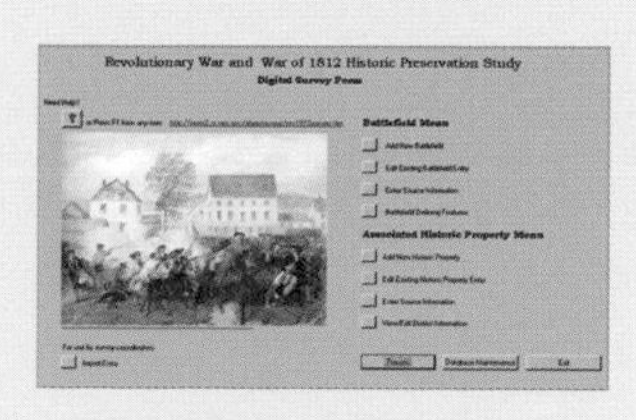

Data Inventory and Review

With over 70 surveyors sending in their work on over 800 sites, a method to allow the CRGIS staff to catalog and review the results had to be created. Using Microsoft Access databases and ColdFusion web database interfaces, CRGIS developed a system with which to inventory and comment on each piece of a surveyor's work, including historical research, spatial data, photography, cartography, and digital survey form.

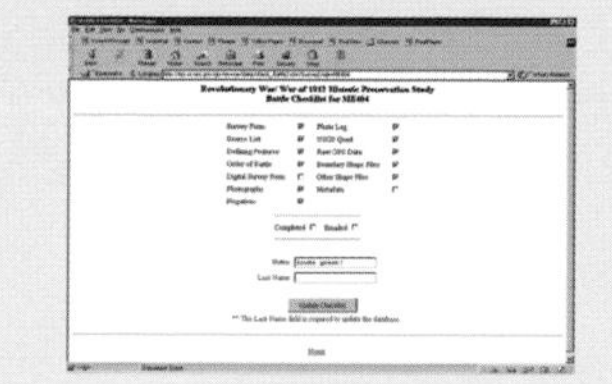

GIS and Remote Sensing

CRGIS will be using the spatial data captured by the surveyors to calculate statistics regarding the condition and threats to surveyed sites. Additionally, CRGIS will run a pilot program using remote sensing to determine if this technology may be useful in identifying historical and cultural landscape patterns. The intent is that by combining these two technologies, better recommendations for preservation and interpretation of these sites can be made.

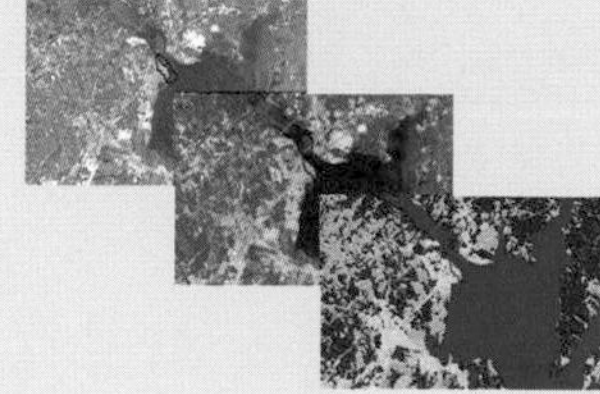

Report to Congress

The culmination of this work will be a report to the U.S. Congress regarding the condition and threats to the Revolutionary War and War of 1812 battlefields and Associated Historic Properties identified by the Advisory Committee. Technology is a vital component in generating the report. GIS will be used to create maps of surveyed sites, and databases will be used to generate forms, charts, and reports that will be central to the interpretation of the results of the study. The use of these technologies will help provide clear findings to Congress in the most expedient and efficient manner possible.

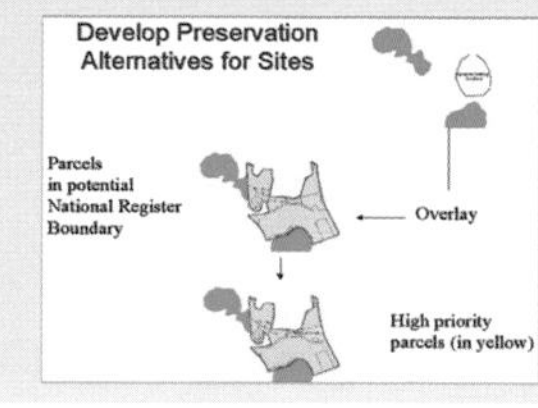

The National Park Service integrated GIS, GPS, remote sensing, and digital database technologies to survey Revolutionary War and War of 1812 sites, as explained in this poster. The study aimed to identify and rank the significance of historic sites, assess threats to these sites, develop preservation alternatives, and offer broad recommendations to preserve and interpret these sites.

Chapter 7

Into the Future: Park Plans

The primary duty of the National Park Service is to protect the national parks and national monuments under its jurisdiction and keep them as nearly in their natural state as this can be done in view of the fact that access to them must be provided in order that they may be used and enjoyed. All other activities of the bureau must be secondary (but not incidental) to this fundamental function relating to care and protection of all areas subject to its control.

Stephen T. Mather, National Park Service director, 1917–1929, February 1925

National parks face daunting challenges from inside and outside their borders in the twenty-first century. Residential growth near our parklands has brought hikers, bicyclists, joggers, and their dogs into sensitive areas. Luxury resorts and cruise ships draw crowds to sensitive coral reefs and fisheries. Pollution clouds scenic vistas and causes acid rain to harm plants, streams, and ecosystems. In some parks, existing facilities are too small or outdated, or were poorly located in the first place. The National Park Service develops long-range plans to navigate these challenges. Today, the park service increasingly relies on a planning tool called GIS, a powerful technology that contributes to data collection, analysis, and communication with the public.

Planning is key to the mission of the National Park Service to protect national parklands for our enjoyment, inspiration, and education. One of the main principles of this mission is making wise decisions based on scientific knowledge. By law, each national preserve needs a general management plan. Each plan maps out a clear vision and framework to guide a national park for the next fifteen to twenty years. The plan serves as a road map to guarantee the survival of the park and its cultural heritage. It outlines how the National Park Service and its partners will reach their goals. Scientific, technical, and scholarly analysis focuses first on the park as a whole, including its regional, national, and global contexts, and then on specific details within the park.

GIS has become indispensable to these plans. For more than thirty years, the park service used the "overlay" techniques of Ian McHarg's *Design With Nature.* Hand-drawn transparent maps of resource data were laid over one another for analysis. Early use of GIS required costly, time-consuming methods to digitize data that many projects could not afford. Now that most parks have digitized basic data, GIS technology has become the tool of choice to create and analyze layers of mapped data. This includes natural and cultural resources, scenic resources, visitor opportunities, and regional land use. GIS technology produces colorful, expressive maps that persuasively illustrate planning ideas to the public and park service managers. It eliminated the old "overlay" technique that took countless hours to revise and reproduce information. GIS saves time and money with its easy manipulation and display of data. The digital data also provides accurate information to measure and evaluate how various plans would affect the park and surrounding area as required in an environmental study. GIS analysis can apply a range of values to resources to help determine which lands may be more sensitive or critical for protection than others. GIS specialists now join park planners to develop alternatives that protect resources while allowing visitors to experience their parks and learn about nature. The evolution of GIS technology offers exciting possibilities as National Park Service planners strive to understand our nation's dynamic ecosystems and protect them for generations to come.

Land Parcel Prioritizing at a New National Park Service Unit

This commemorative marker sits on a bluff overlooking the site of the Sand Creek Massacre.

On November 29, 1864, U.S. soldiers attacked a village of about five hundred Cheyenne and Arapaho people camped along the banks of Big Sandy Creek in southeastern Colorado. The troops massacred some 150 people, mostly women, children, and the elderly, who believed the U.S. Army would protect them. The government eventually condemned the killings. In November 2000, federal law authorized Sand Creek Massacre National Historic Site to recognize its significance in American history and importance to Cheyenne and Arapaho peoples, including descendents of the massacred. The National Park Service acquired 880 of 12,583 acres within the authorized boundary. The site will not open to the public until the park service and its partners acquire enough private and state land to ensure its preservation. GIS technology will help make that a reality. GIS data helped pinpoint where the tragedy occurred. GIS mapping also shows the geographic relationship between the artifact sites, Big Sandy Creek, and views of the area from six different points. GIS specialists have updated the boundaries and status of land ownership, and the tribes' oral history of the event identified significant areas. The park service added these new information layers, or themes, to the GIS model. A computer analysis of the model rated the various lands based on their historic value. This has helped the park service and its partners decide which lands to pursue, ultimately to establish a national historic site for our benefit and education.

In 1864, U.S. soldiers attacked a village of Cheyenne and Arapaho people camped along the banks of Big Sandy Creek in southeastern Colorado.

Robert Lindneaux's painting from the 1930s or 1940s depicts the massacre. Courtesy of the Colorado Historical Society.

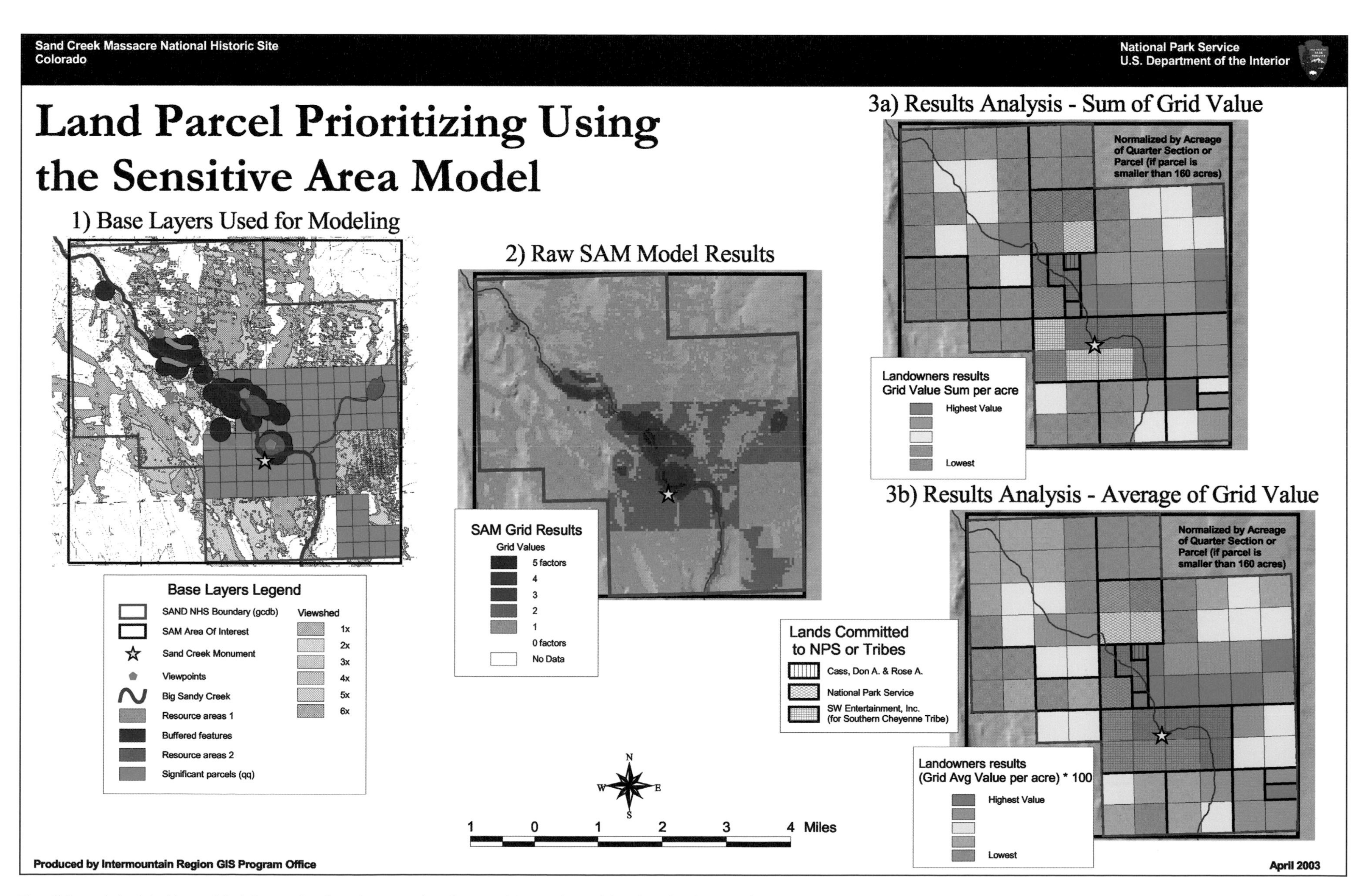

This GIS map helped the National Park Service identify and rate parcels in the area of Big Sandy Creek based on their historic value.

GIS Visualization, Alternatives Development, and Impact Analysis

Lincoln Bridge at Travertine Creek is a popular feature at the national recreation area, at right. The park's natural resources include Antelope Springs, below, and Buffalo Springs, below at right.

Native Americans called it the "Peaceful Valley of Rippling Waters." Today, Chickasaw National Recreation Area draws about 3.4 million visitors a year to its natural springs, lakes, trails, and geological and hydrological features. Covering about ten thousand acres in south-central Oklahoma, the recreation area lies in a transition zone where eastern deciduous forests meet western prairies. It also preserves classic examples of Civilian Conservation Corps architectural craftsmanship and ingenuity from the public works era of the 1930s. Created as Platt National Park in 1906 and renamed in 1976 to honor the Chickasaw Nation, the recreation area relies on GIS to help direct its management policies in the twenty-first century. The park service, along with the assistance of CommunityViz, a program of the nonprofit Orton Family Foundation, is creating a general management plan to guide resources, visitor use, interpretation, and recreation facilities for the next fifteen to twenty years. CommunityViz serves as a tool to build public involvement, illustrate alternatives, and analyze impacts of management alternatives. Its software created a 3-D display of the park for use at public meetings. The interactive model helped show the current state of the park, its resources, and its relationships with surrounding communities and lands. The park service used CommunityViz software to create GIS models and maps and three separate 3-D zoning proposals, including a virtual fly-through of each. CommunityViz also helped analyze the environmental impacts of the various alternatives. This will result in a new general management plan for the recreation area based on a fuller understanding of alternatives and their effects on the region.

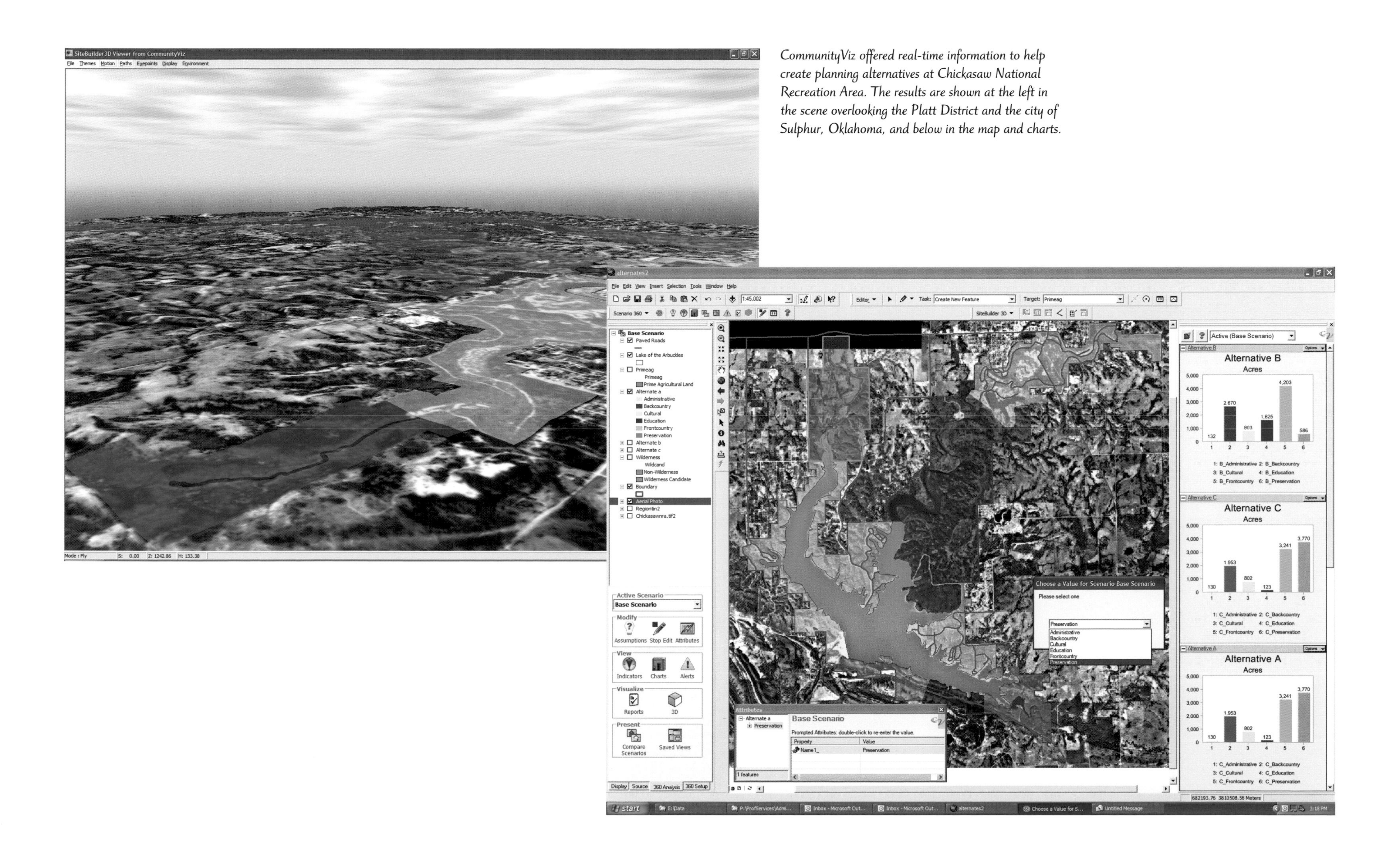

CommunityViz offered real-time information to help create planning alternatives at Chickasaw National Recreation Area. The results are shown at the left in the scene overlooking the Platt District and the city of Sulphur, Oklahoma, and below in the map and charts.

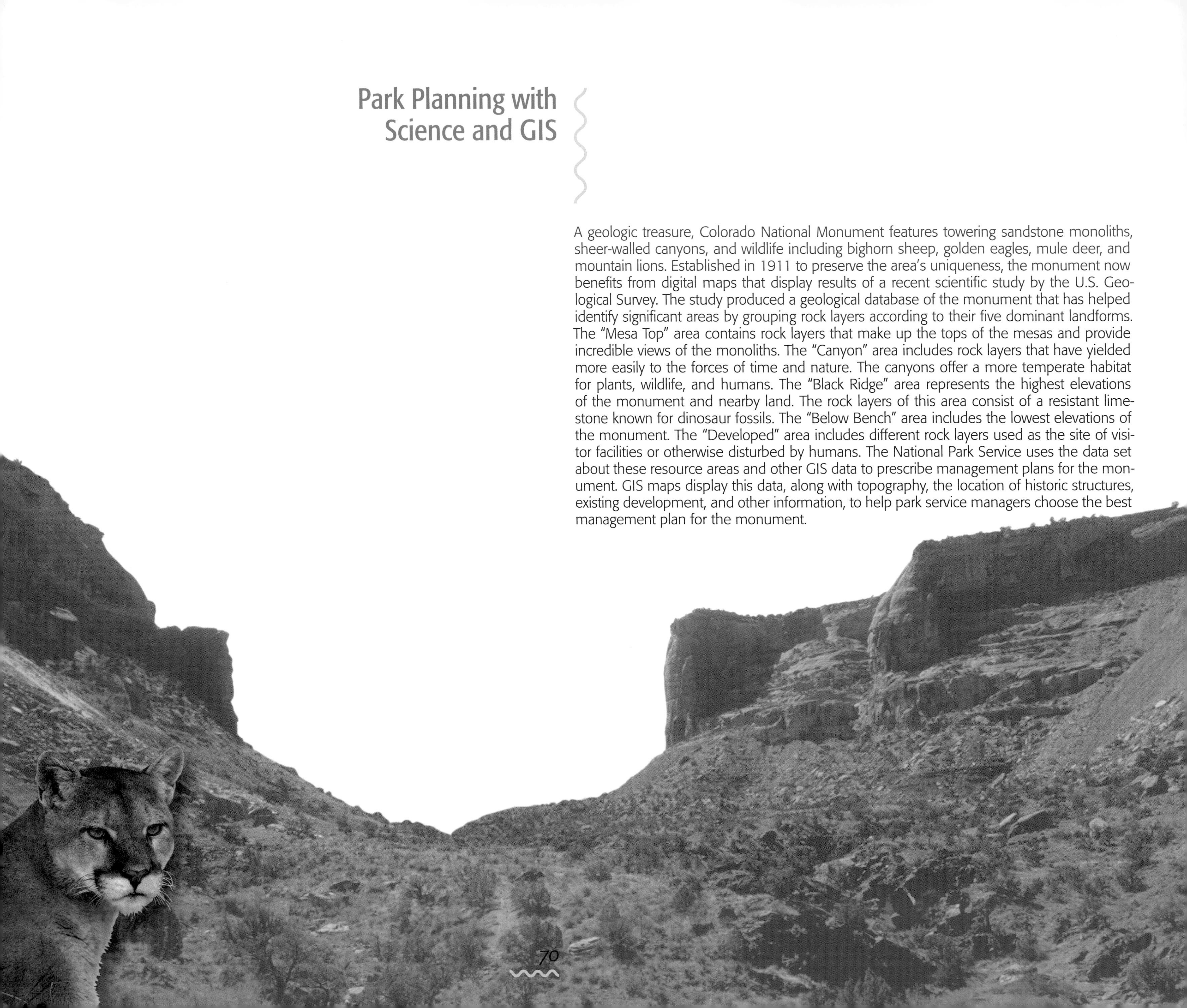

Park Planning with Science and GIS

A geologic treasure, Colorado National Monument features towering sandstone monoliths, sheer-walled canyons, and wildlife including bighorn sheep, golden eagles, mule deer, and mountain lions. Established in 1911 to preserve the area's uniqueness, the monument now benefits from digital maps that display results of a recent scientific study by the U.S. Geological Survey. The study produced a geological database of the monument that has helped identify significant areas by grouping rock layers according to their five dominant landforms. The "Mesa Top" area contains rock layers that make up the tops of the mesas and provide incredible views of the monoliths. The "Canyon" area includes rock layers that have yielded more easily to the forces of time and nature. The canyons offer a more temperate habitat for plants, wildlife, and humans. The "Black Ridge" area represents the highest elevations of the monument and nearby land. The rock layers of this area consist of a resistant limestone known for dinosaur fossils. The "Below Bench" area includes the lowest elevations of the monument. The "Developed" area includes different rock layers used as the site of visitor facilities or otherwise disturbed by humans. The National Park Service uses the data set about these resource areas and other GIS data to prescribe management plans for the monument. GIS maps display this data, along with topography, the location of historic structures, existing development, and other information, to help park service managers choose the best management plan for the monument.

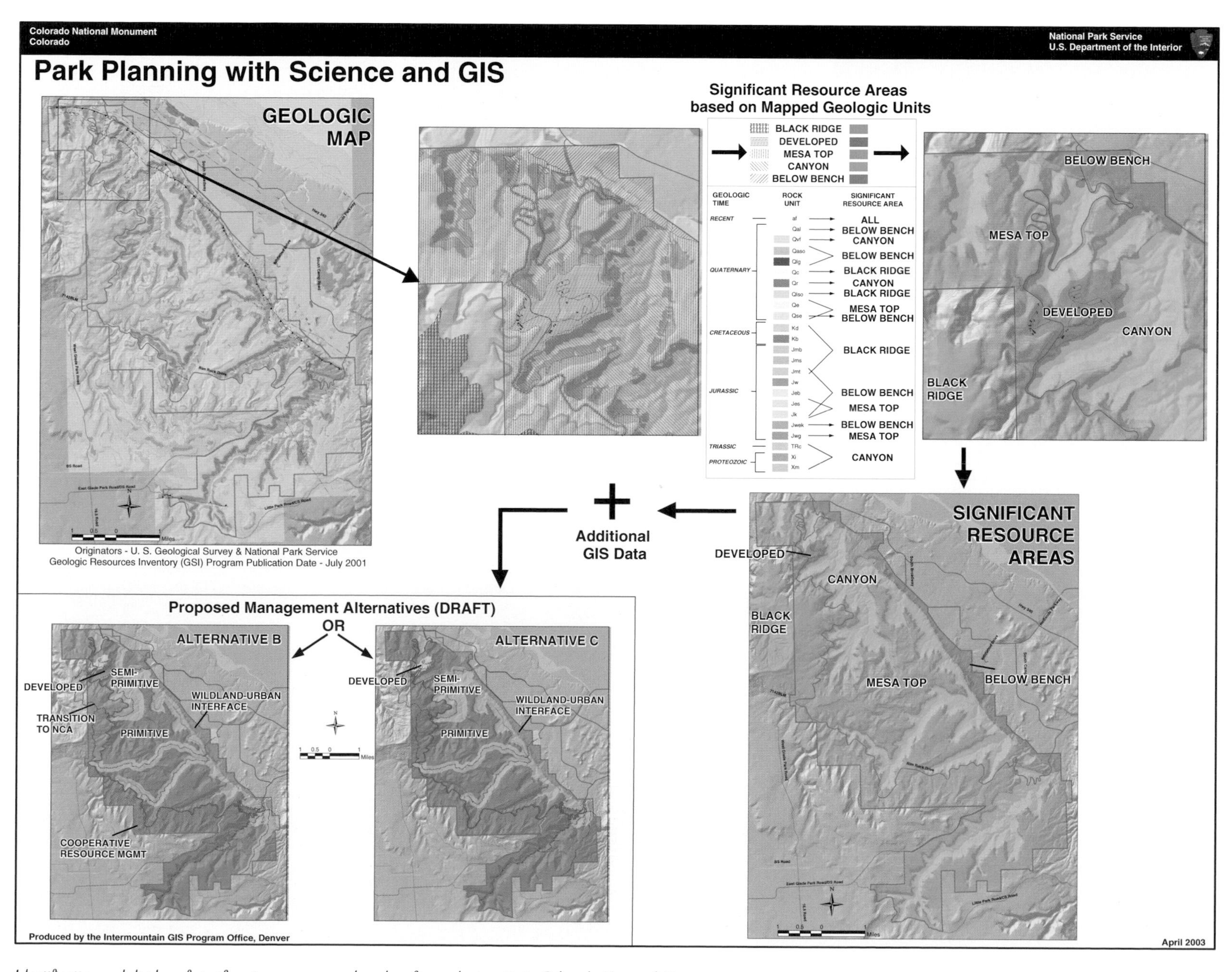

Identification and display of significant resource areas based on five geologic units in Colorado National Monument.

GIS in the General Management Plan

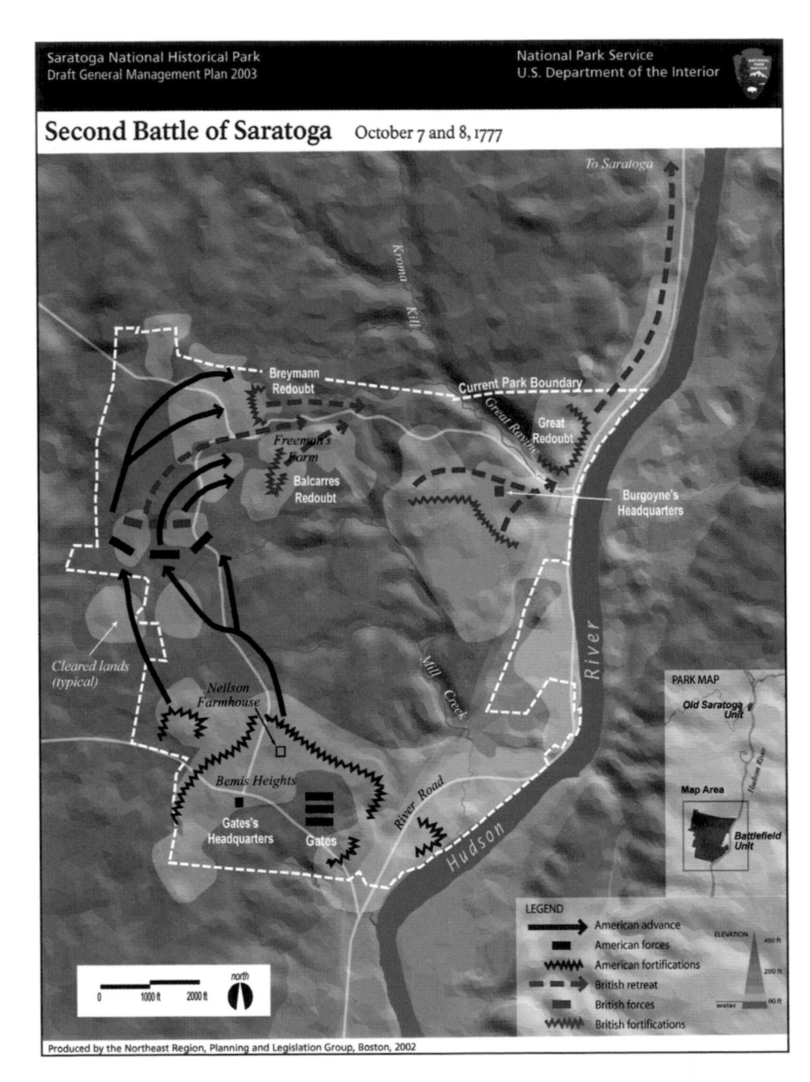

Historic troop positions and movements of the second battle of Saratoga are shown with the current park boundary and topography to give a clear sense of what happened.

The dramatic topography of steep slopes, ravines, and valleys, and the mosaic of fields and forests played a key role in the first significant American victory of the Revolutionary War. The battlefields where American and British armies met in 1777 are now part of Saratoga National Historical Park in New York. Here, a major British army surrendered to American forces, leading France to recognize the independence of the United States and enter the war as a decisive military ally. Today, GIS technology helps explain the American victory, which proved to be a major turning point in the war. The British army conceived a plan to quell the rebellion with a single decisive military campaign. The strategy hinged on the invading army dividing the upstart colonies along a natural corridor of rivers and lakes stretching from Canada to New York City. The Americans chose an area of the upper Hudson River near Saratoga to fight back, relying partly on the hills, river, and heavy forest to slow the enemy along the riverbank. The forbidding geography and position of American patriots forced the British to choose between a frontal attack between hills and the river or a flanking maneuver through forested uplands. The British tried to flank the Americans and failed.

The combination of digital elevation information and color aerial photography becomes a 3-D tool to help park interpreters orient visitors to the varied terrain and features of the battlefield, and to show the battlefield's relationship to the Hudson River.

The park service today uses a GIS map to help visitors understand the battles. The map combines aerial photography with elevation data and uses GIS tools to display light and shadow on terrain from a given angle of the sun. The map display of the varied topography illustrates the difficulties the British faced at Saratoga. The park service also created a series of GIS maps to help visitors understand the events leading up to the British surrender. The maps display the historic fields and forests, roads, troop movements, and topography in relationship to the current park boundary. GIS also plays a planning role to restore key landscapes as they existed in 1777 and create viewing areas to show what the opposing armies faced and how they took advantage of landscapes to serve tactical needs. A GIS analysis compared four alternatives for viewing areas and to configure fields and forests as they were at the time of the battles. GIS technology showed how each alternative would change the landscape and displayed each one on a map to help the park service choose the best option. Overall, GIS helped managers in many ways as they developed a general plan to guide the park's future.

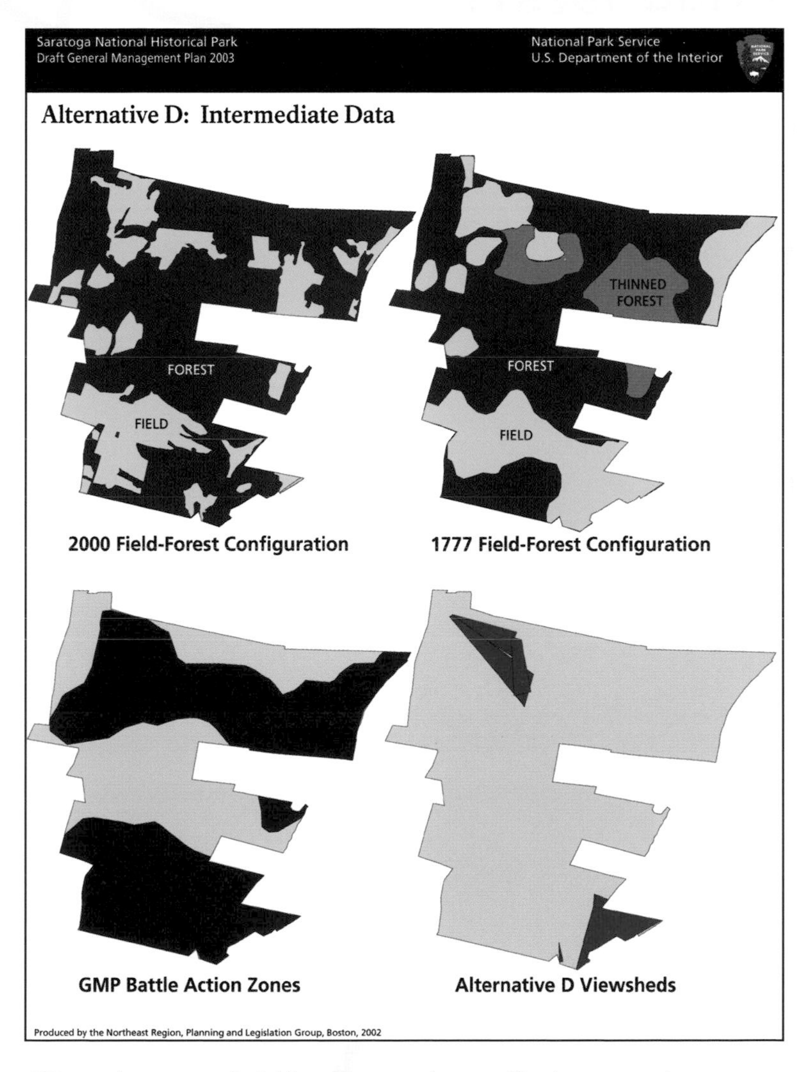

GIS is used to compare the fields and forests at the time of battle in 1777 to the configuration in 2000 and illustrates battle action zones and viewsheds. These maps represent the four sets of data used in the field–forest analysis: existing field–forest as of 2000, visitor viewsheds, historic field–forest, and the battle action zones. This image is part of a series that displays proposed changes to the fields and forests at the battlefield. Changes, such as clearing, thinning, and planting, were proposed in certain areas to more accurately reflect how the landscape appeared during the battles, but also considered visitor needs and sensitive natural environments.

The restored Neilson home looks much as it did when American staff officers used it for quarters during the battles of Saratoga. The cannon is a replica of the type Americans used during the battles.

Above, a cannon overlooks the Hudson River, where American forces occupied high ground, forcing the British to leave the river road and fight in more difficult terrain. Below, reenactors re-create the battles of Saratoga for visitors.

Mapping Proposed Park Expansions

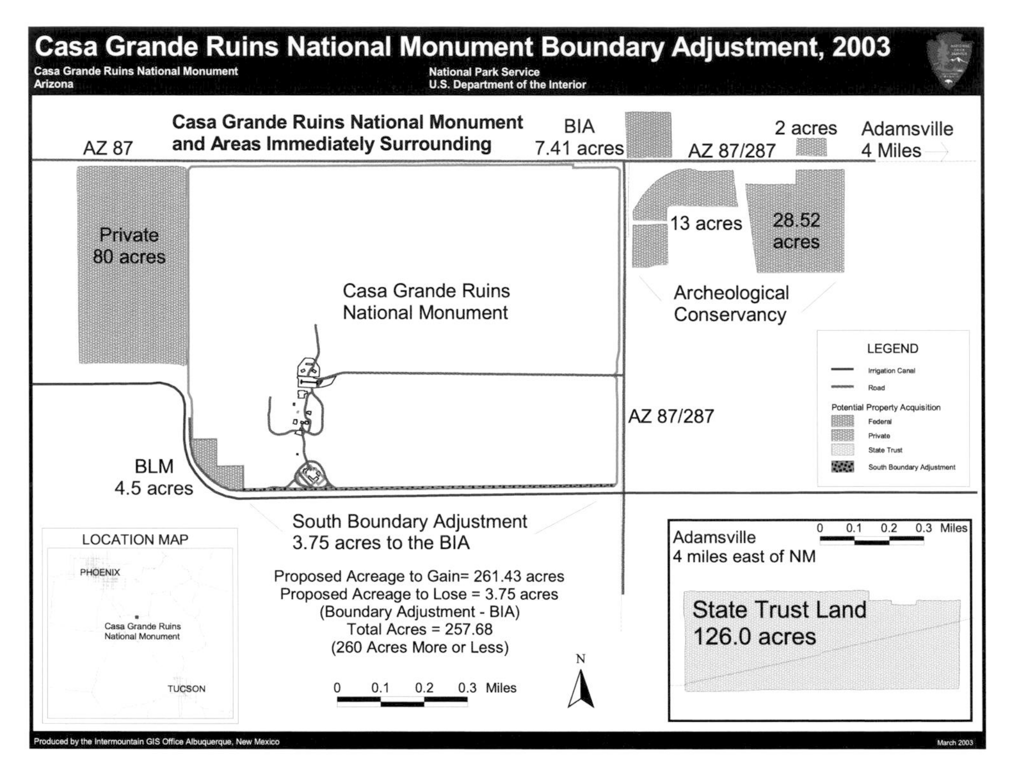

GIS illustrates acreage gains and losses in a recent Casa Grande Ruins National Monument boundary adjustment.

The nation's first cultural preserve and a national park filled with petrified wood are experiencing growing pains in Arizona. The National Park Service wants to protect archaeological sites, fossils, wildlife habitat, and scenic vistas at Casa Grande Ruins National Monument and Petrified Forest National Park. The park service and its partners used GIS technology to display proposed expansions of the parks on digital and paper maps for easy review.

Established as the nation's first cultural preserve in 1892, the Casa Grande monument protects a mysterious prehistoric ruin known as the "Big House" and more than sixty known archaeological sites near the fast-growing community of Coolidge. As a result of recent testing and excavation projects, the park service now knows a great deal more about the nature of archaeological remains on the outskirts of the Casa Grande ruin. The park service would like to add some 260 acres to the 472.5-acre monument to protect hundreds of ancient pit houses, burial sites, canals, and other deposits buried in outlying areas that have been farmed for decades. The expansion would protect archaeological sites, many of them threatened by proposed commercial development on the north side of Coolidge.

First established as a monument in 1906, the Petrified Forest was accorded national park status in 1962. It features one of the world's largest and most colorful concentrations of petrified wood, spectacular views of the Painted Desert, diverse archaeological sites, and wildlife including rattlesnakes, bats, and owls. Since the last expansion, paleontologists have found that an escarpment bisecting the park contains the best record of Triassic-era ecosystems found anywhere in the world. Hundreds of archaeological sites also have been identified near current park boundaries, including pueblo ruins and some of the most unusual rock art panels in the Southwest. Scientists have stated that the resources located just outside the park are likely more valuable than those currently protected. Most of the scientifically valuable area is undeveloped ranch land, which has helped preserve scenic views. But a proposed landfill, subdivision of some sections for small ranches, mechanized mining of petrified wood on private and state lands, and other plans threaten scenic views and resources. In one instance, private investigators caught a resident using a backhoe and flatbed truck to excavate a 75-foot-long petrified log that would have been worth several hundred thousand dollars when cut into polished tabletops. The proposed expansion would more than double the size of the 93,533-acre park, partly to prevent pottery theft, rampant vandalism, illegal fossil removal, and imminent development that could harm archaeological sites, scenic views, and prehistoric trees turned to stone.

CHINDE MESA

Pilot Rock
6234ft
1900m

BLACK FOREST

Land Ownership

Petrified Forest National Park

Legend

NPCA Proposed Additions

Private Ownership
Twin Buttes Ranch LLC
Hatch
McCauley, Hatch, Muse
Bob Worsley/NZ Corp. Properties
Jeffers
Multiple Private Owners

Other Ownership
NPS, Petrified Forest NP
BLM
State of Arizona

N

0 2 4 Miles

Produced by the Intermountain GIS Office Albuquerque, New Mexico

Revised 8/13/03

Color coding helps identify land ownership in and around Petrified Forest National Park.

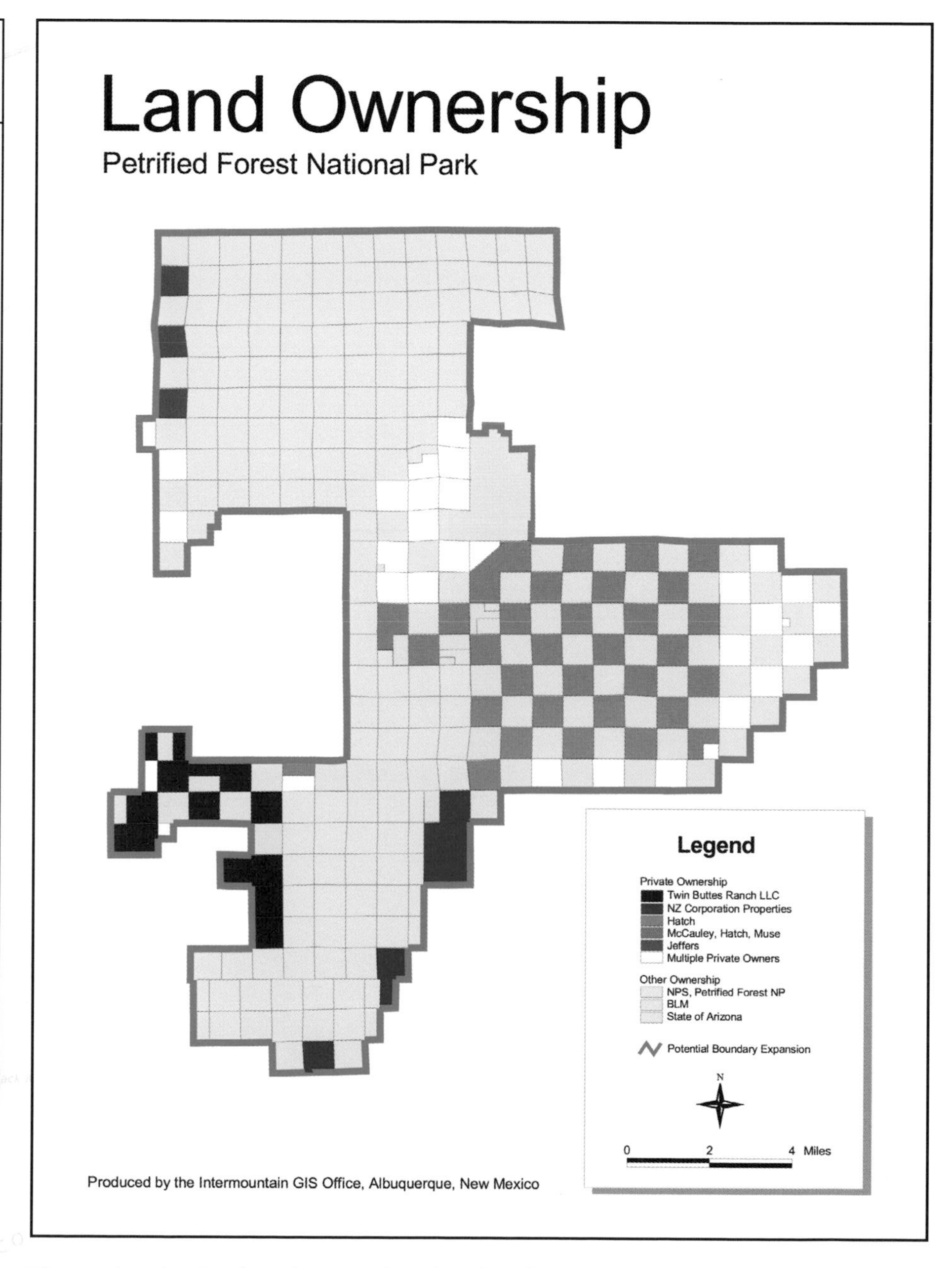

The green boundary line shows the potential new boundary of an expanded Petrified Forest National Park.

Crystal Forest

Rainbow Forest

Alternative Management Zones for Monocacy National Battlefield

The battle of Monocacy in the context of surrounding communities and transportation systems.

Known as the "Battle That Saved Washington," the battle of Monocacy on July 9, 1864, between some 15,000 Confederate soldiers and 5,800 Union troops marked the last campaign of the Confederacy to carry the Civil War into the north. The Confederates won the battle, fought on a checkerboard of wheat and cornfields in Maryland. But the fighting cost them a day's march and their last chance to capture Washington, D.C., just thirty-five miles distant. The battle left some 2,000 soldiers dead or wounded. Later, Confederate General Jubal Early stood inside the District of Columbia, looking at the earthworks of Fort Stevens as Yankee reinforcements arrived to defend their capital against an exhausted foe. Early's forces soon retreated across the Potomac River to friendly Virginia.

To preserve, understand, and explain the importance of this historic battlefield, the National Park Service is using GIS to display various options for five management zones in the park. Displaying these zones on maps helps park managers focus on the best ways to preserve historic buildings and landscapes, maintain natural resources, commemorate battlefield events, and provide visitor, administrative, and maintenance facilities. The park service turned its hand-drawn map overlays of these zones into digitized maps.

Left: The Thomas farm is one of five historical farms within Monocacy National Battlefield.

Center: This monument honors soldiers from New Jersey who fought in the battle.

Right: Civil War–period artillery piece at the battlefield.

The area is displayed in the context of its importance as a military and transportation crossroads during the Civil War.

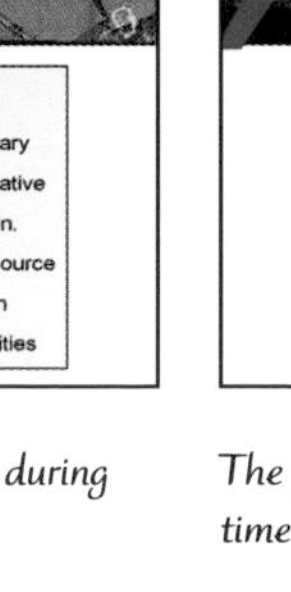

The five management zones of Monocacy are shown in relation to the story of the local communities at the time of battle.

Potomac Heritage National Scenic Trail

Citizens in the Potomac Heritage National Scenic Trail corridor are rediscovering history and reclaiming access to rivers and other outdoor places. Communities in Maryland, Virginia, Pennsylvania, and the District of Columbia are using the designation of the trail in 1983 to develop and connect trails, historic sites, and a range of recreational and educational opportunities. The scenic trail specifically seeks to connect the Chesapeake Bay and the Laurel Highlands of western Pennsylvania with a series of trails. To display possible trail links, the National Park Service and its partners used GIS software to map the location of existing, planned, and potential trails and their geographic relationship to park service and state lands. Unlike most national parks, the trail corridor includes parts of two national parks, Chesapeake & Ohio Canal National Historical Park and George Washington Memorial Parkway. Linking the trails in this area with new segments requires a partnership between the park service and state and local governments. The scenic trail recognizes the Laurel Highlands Hiking Trail, the Chesapeake & Ohio Canal towpath, and the Mount Vernon Trail, and has plans to recognize more. The 70-mile Laurel Highlands trail runs along Laurel Ridge between the Youghiogheny River and Johnstown, Pennsylvania. The trail features overnight shelters and tent-camping areas. The 184.5-mile Chesapeake & Ohio Canal towpath runs along the Potomac River between Georgetown in Washington, D.C., and Cumberland, Maryland. Highlights include a series of locks and aqueducts, and remains of a canal that operated from 1828 to 1924. The 18.5-mile Mount Vernon walking and biking trail offers views of Washington, D.C., and opportunities to visit national memorials as it parallels the Potomac and George Washington Memorial Parkway between Theodore Roosevelt Island and Mount Vernon.

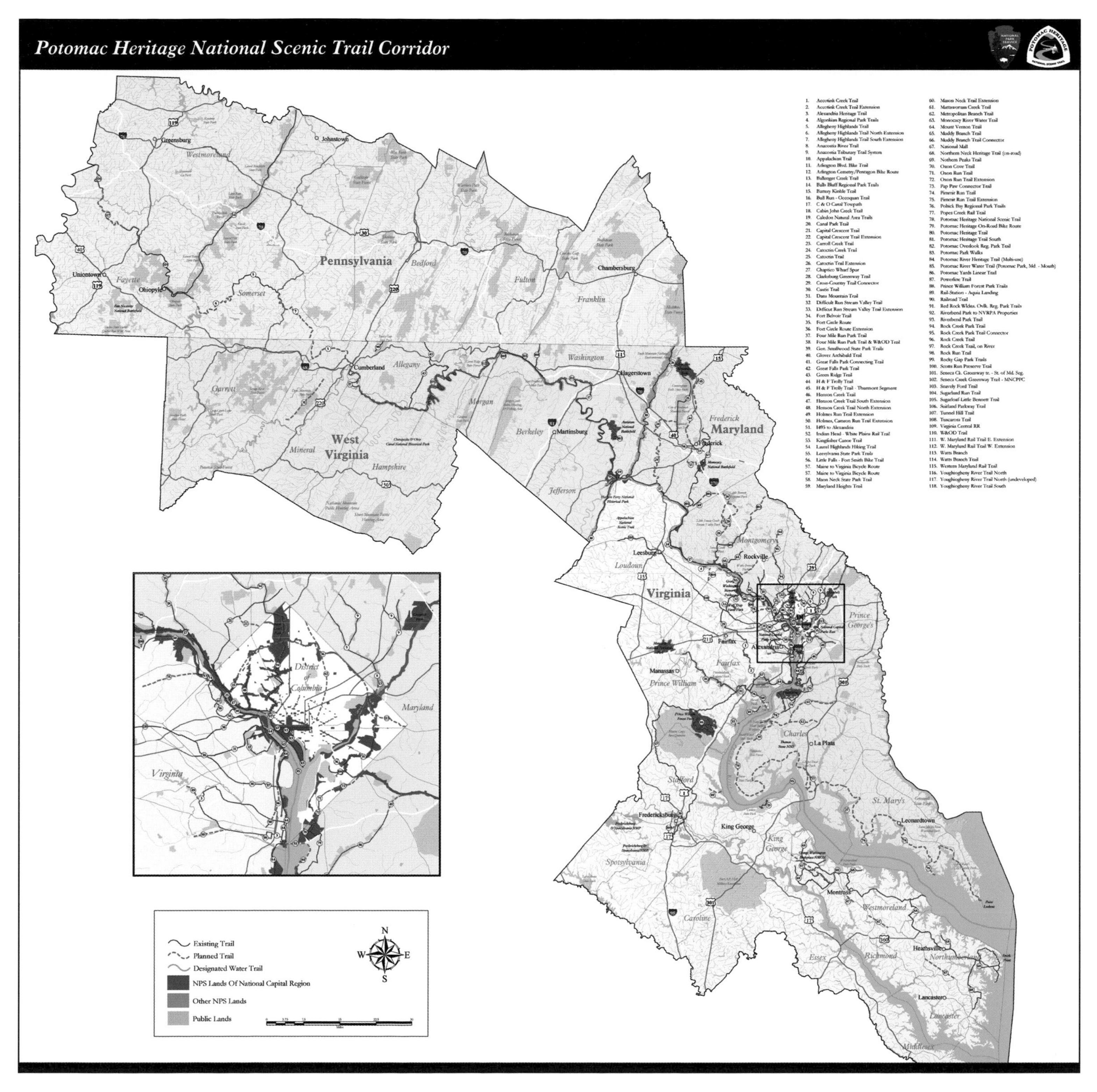

This GIS map illustrates the need for a partnership between the park service and state and local governments to make the goal of a trail corridor a reality.

Chapter 8

Safe and Sound: Roads and Trails

If more Americans could be induced to visit these scenic treasure houses the public will come to appreciate their value and stand firmly in their defense.

Allen Chamberlain, Appalachian Mountain Club, 1911

It's often the things we take for granted that make our national park vacations so memorable. We may choose a park for its natural features, such as geysers or glaciers, or simply because we want a break from the hustle and bustle of everyday life. As we enjoy nature's offerings, we may not notice the smooth roads, working restrooms, and safe trails that help form the backbone of the national park system. We tend to take our safety and personal comfort for granted before we set off from a visitor center in search of a bald eagle, or a distant mud pot. Imagine our reaction, then, if we arrived to find a park full of potholed roads, dangerous trails, broken toilets, and splintered benches. To make sure that doesn't happen, national parks use all the tools at their disposal, including the latest in GIS software and other computer programs.

National parks often go to great lengths to provide basic services such as drinking water and sewers. Some parks capture millions of gallons of rainwater. Others build sophisticated water-recycling plants to conserve what little they have. Parks must restore trails and repair buildings in rugged, remote areas far from civilization. Many buildings are more than fifty years old and require special care to protect their architectural and historical significance. To meet these challenges, national parks increasingly use GIS to map the locations of roads, trails, utilities, and buildings along with their condition and need for repair. GIS technology turns reams of complex data into a clear map with many layers, each showing one or more themes. Using GIS maps, park managers can identify and schedule needed repairs and better understand the work and cost involved. In this way, GIS serves the national park mission to protect parklands for our enjoyment and education. And when we don't have to worry about the condition of roads, trails, and restrooms, we can focus on all the things that make our national park experience so special.

Utility Mapping in the Intermountain Region

Park utilities at Aztec Ruins National Monument in New Mexico range from flagpoles and fire hydrants to water lines and parking lots.

The National Park Service wants to capture the knowledge of its veteran maintenance crews and managers by recording what they know about the position of water and sewer lines and other utilities across our national parklands. In the past, people who oversaw park roads, buildings, trails, and campgrounds kept most of the information on paper, or in their heads. Now many of those managers are retiring without leaving current maps. To preserve this valuable knowledge, park service teams are using GPS and GIS technology, first to record the information and then to create maps that show the locations of park utilities. GPS systems accurately record utility locations to one meter, and interviewers can record a worker's comments about things like pipe length, width, and date of installation or repair. Maintenance crews check resulting GIS maps for accuracy and make additional notes as needed. Examples of success so far include Aztec Ruins National Monument, Fort Union National Monument, El Morro National Monument, and Salinas Pueblo Missions National Monument in New Mexico; Casa Grande Ruins National Monument in Arizona; and Hovenweep National Monument in Utah.

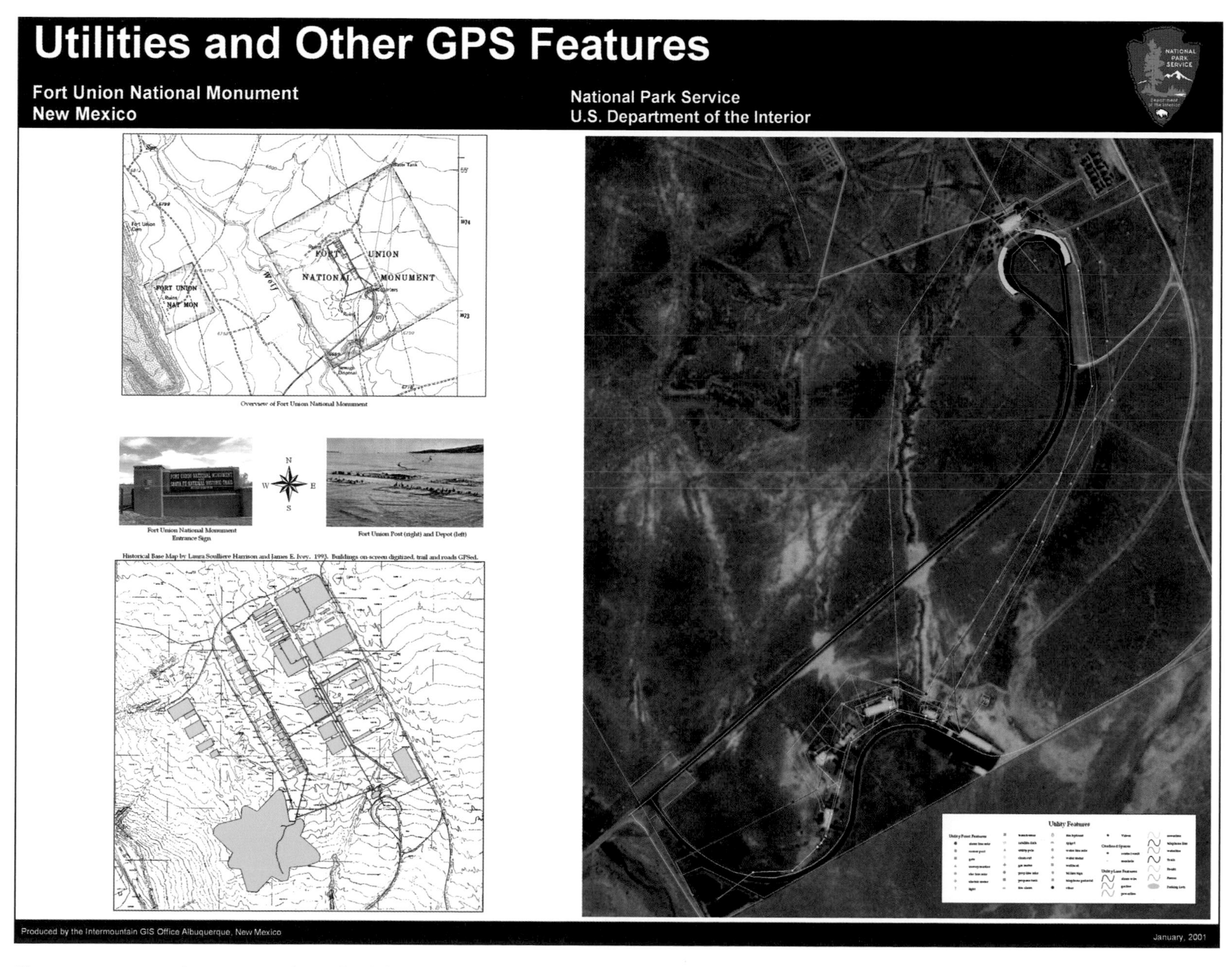

Electric meters, survey markers, propane tanks, satellite dishes, and other utility features are identified on this poster that also includes a historical basemap, photographs, and monument overview.

Using the Power of a GIS Trail Inventory for Trail Planning

The coastal mountains near Los Angeles drop sharply to the Pacific Ocean in a patchwork of public and private lands that make up the Santa Monica Mountains National Recreation Area. Federal, state, and local park agencies cooperate with private preserves and landowners to protect a diverse Mediterranean ecosystem where native peoples once made their homes. Now, the national recreation area is "L.A.'s backyard" for hiking, mountain biking, walking dogs, bird watching, and horseback riding along a network of 320 miles of trails and dirt roads. Here, GIS serves as a trail guide of sorts for the National Park Service and its partners as they plan the future of the new sixty-mile Backbone

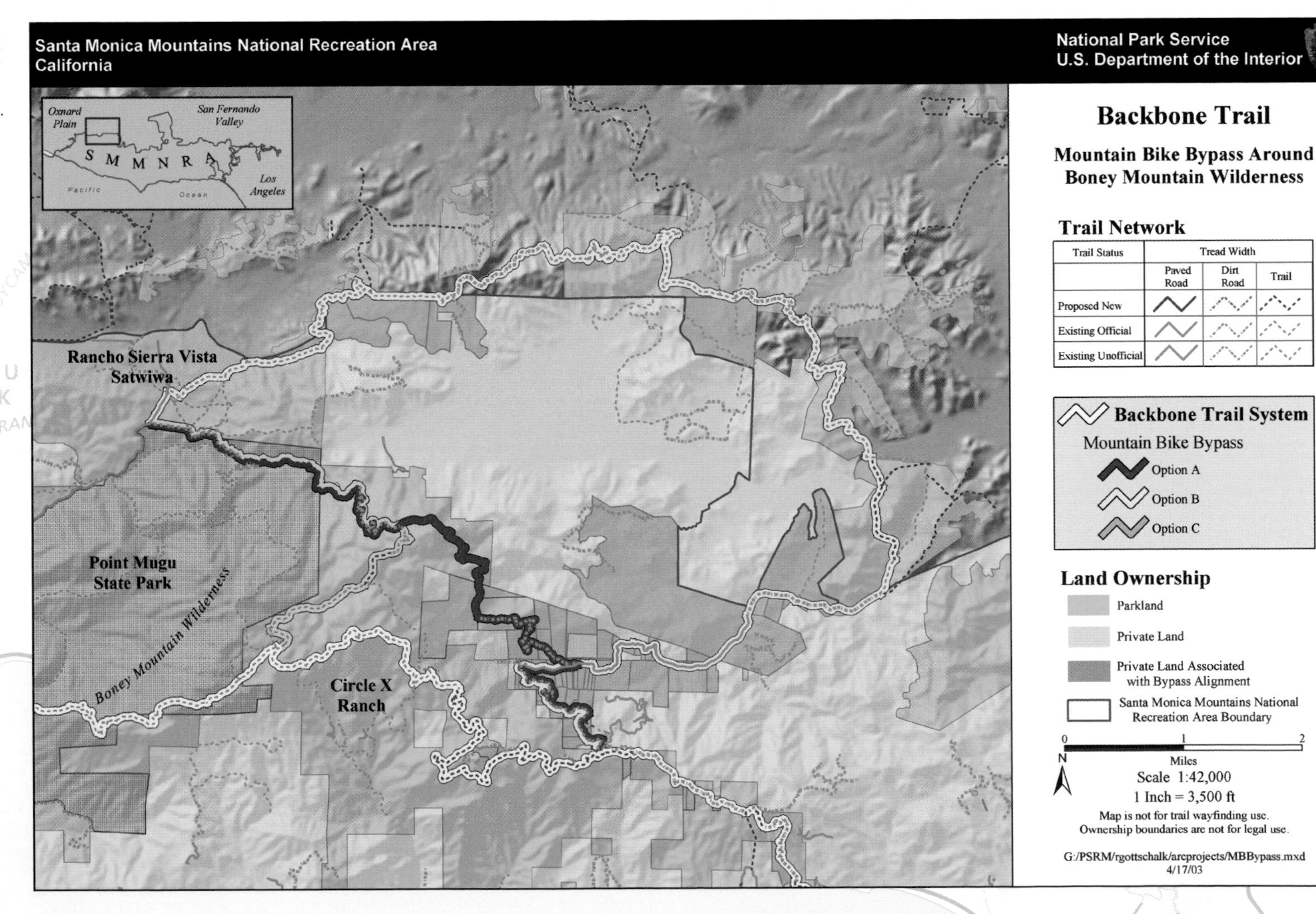

Options for a mountain bike bypass around the Boney Mountain Wilderness are illustrated on the Backbone Trail in the Santa Monica Mountains National Recreation Area.

Trail. Planners studied the trail conditions and plotted the information onto GIS maps. The maps quickly displayed patterns that might otherwise have gone unnoticed in reams of data. In this case, GIS maps showed the risk of erosion based on steep trail grade in some places. The information will help trail planners decide whether to limit activity on the trail or even reroute certain sections. GIS software produced maps that illustrate the geographic relationship between key aspects, or attributes, of the trail. The map of the Backbone Trail showed at least ten attributes, including trail width, trail activities, alternate routes, and future planning options to balance trail use with preservation.

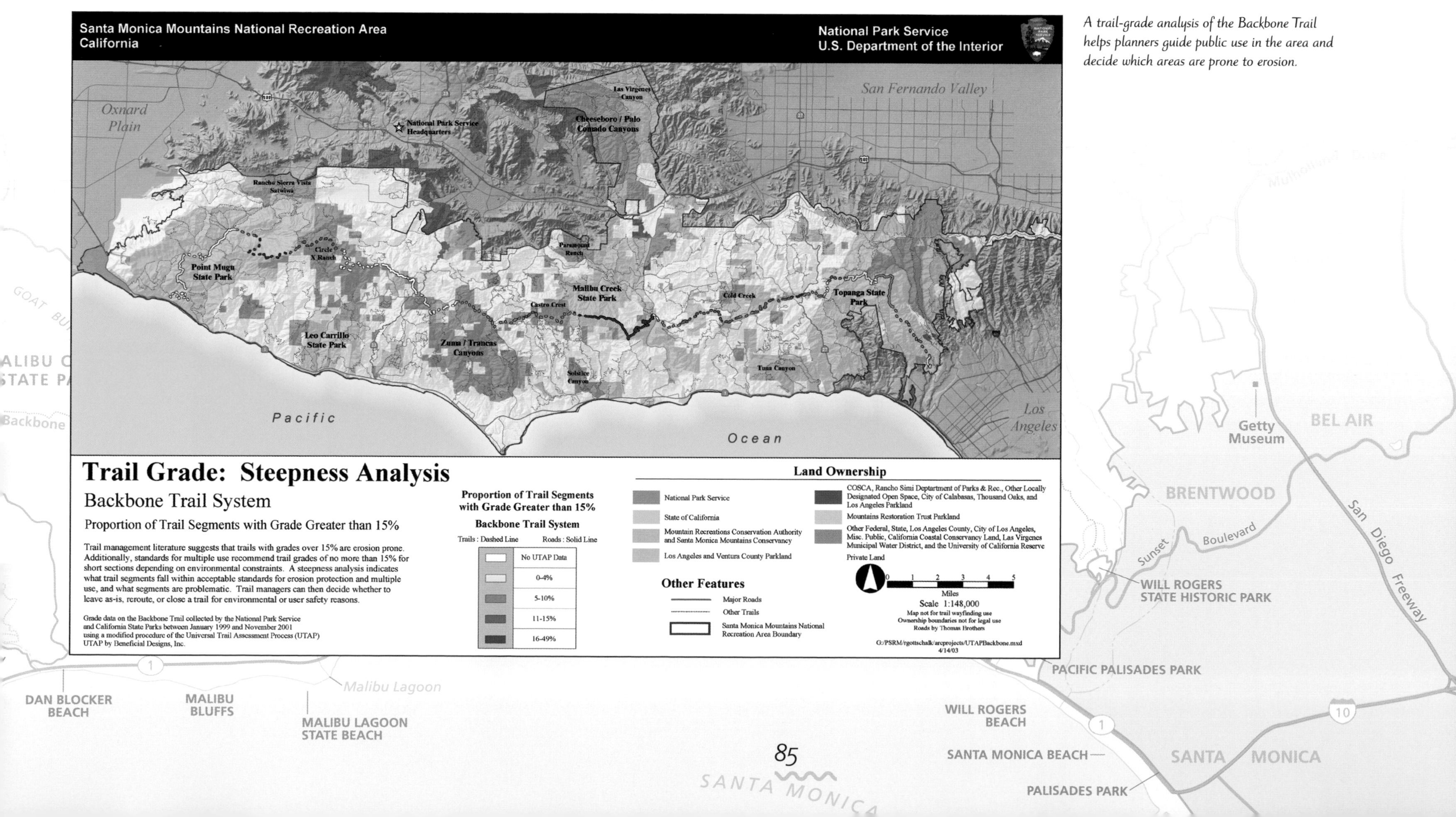

A trail-grade analysis of the Backbone Trail helps planners guide public use in the area and decide which areas are prone to erosion.

GIS Database for Rehabilitation of Going-to-the-Sun Road

National Park Service employee Kerri Mich uses a GPS receiver to collect submeter GPS positions to map culverts along Going-to-the-Sun Road in Glacier National Park.

Soaring peaks, lush forests, deep-blue lakes, glaciers, and sculpted mountains make Going-to-the-Sun Road a spectacular drive by anyone's standards. Carved from Montana mountainsides for part of its fifty-mile length in Glacier National Park, the road is an engineering marvel threatened by age, weather, poor drainage, and deterioration. The road opened in 1933 and needs extensive repairs to avoid the risk of catastrophic failure. The National Park Service turned to GIS to help restore the road for modern travel without marring the surrounding natural, cultural, and historical resources. The park service created a database to

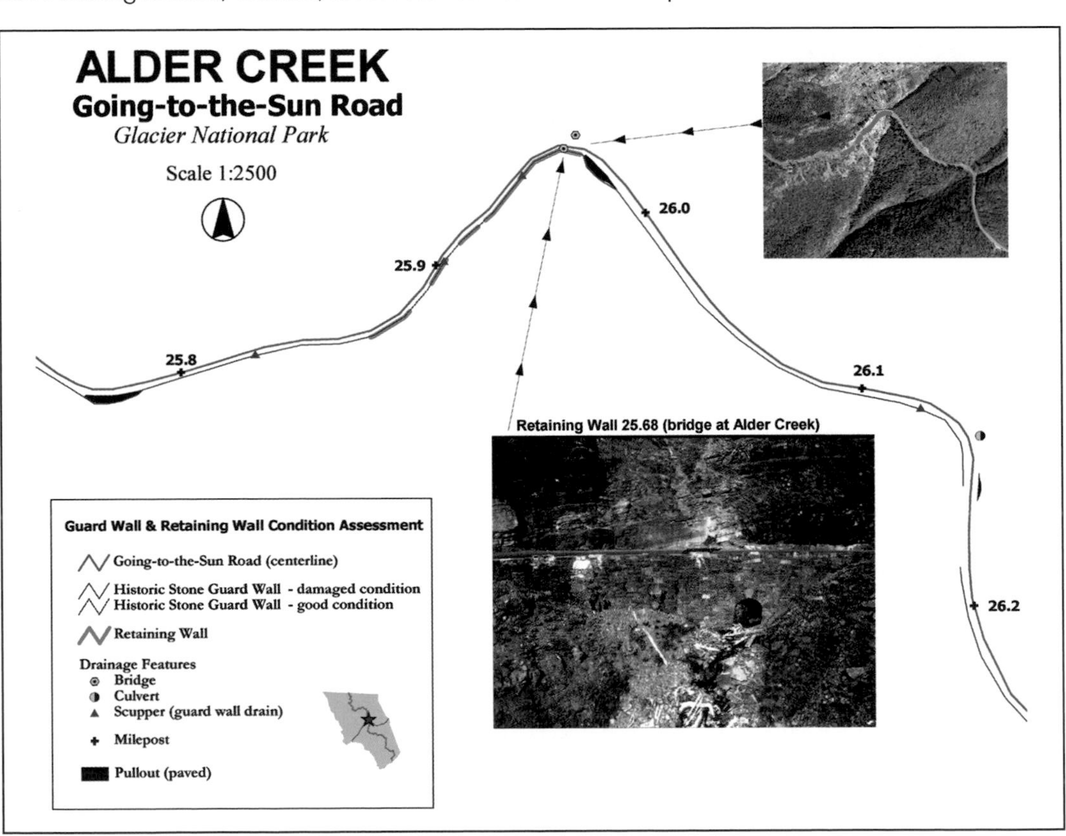

Drainage and road features, guard walls, and pullouts along Going-to-the-Sun Road, with a photograph of the bridge and culvert at Alder Creek in Glacier National Park. GPS helped map road features where Alder Creek intersects Going-to-the-Sun Road.

display themes of the road on a GIS map. The themes, or map layers, identified mileposts, pullouts, guard walls, drainage culverts, survey points, wayside exhibits, avalanche chutes, trailheads, retaining walls, and intersecting roads. The park service linked this information to other databases that described the condition of the road and historic guard wall. Developing a GIS database with help from GPS software will provide a consistent map to help engineer and design repairs and identify cultural and historic items for preservation.

This aerial photograph of the Logan Pass area of Going-to-the-Sun Road is overlaid with information that illustrates road and drainage features. Logan Pass is the apex of Going-to-the-Sun Road and a key attraction for most park visitors. It demands a great deal of road-management attention because of the historic guard wall's deteriorating condition caused by avalanches and drainage issues. The image shows the severity of the terrain and the abundance of mapped drainage features, while also showing the capacity for visitor use.

Going-to-the-Sun Road winds through Glacier National Park.

Habitat Analysis and Road Relocation following Hurricane Opal

Evidence of Hurricane Opal's fury at Gulf Islands National Seashore.

Hurricane Opal demonstrated the power of nature when the storm hit Florida's Santa Rosa Island in 1995. Sustained winds of 125 miles per hour and a 15-foot storm surge heavily damaged buildings, destroyed roadways, eroded shoreline, razed sand dunes, and filled wetlands with sand in this part of Gulf Islands National Seashore. To help document damage and rebuild the park, the National Park Service used GIS software and global positioning system technology to create maps. GIS helped show how the island literally "rolled over" on itself in the storm, moving 200 feet closer to the mainland in certain areas, with the gulf beach receding by as much as 150 feet. The hurricane gave the park service the perfect opportunity to reevaluate the island's main road, which had blocked the natural growth and migration of the dunes, fragmented their habitat, and disrupted a second dune system. GIS mapping illustrated the problem and helped convince the Federal Highway Administration to relocate three miles of the seven-mile-long road. The park is also using GIS to help identify areas prone to storm flooding and damage, and island features to avoid during future reconstruction.

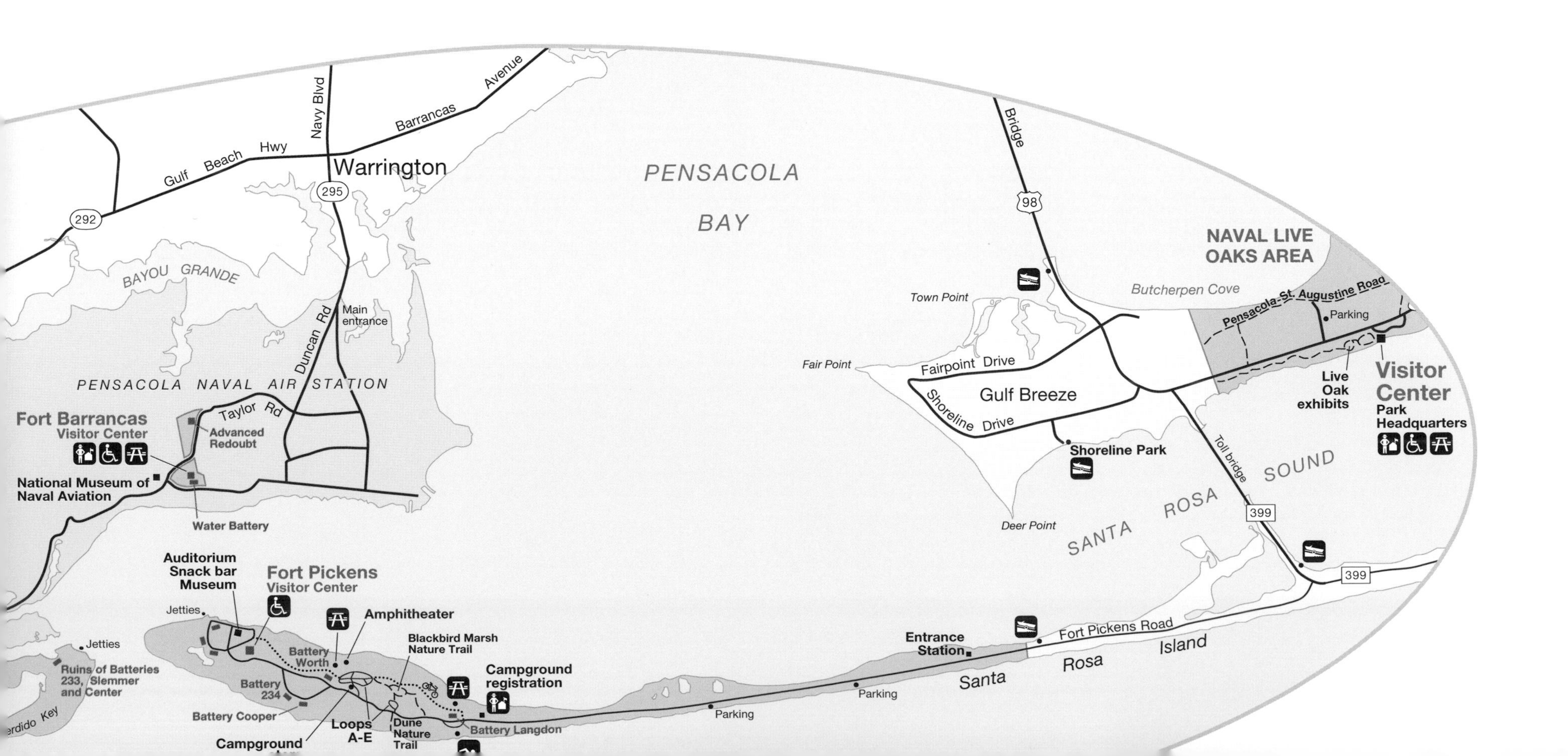

display themes of the road on a GIS map. The themes, or map layers, identified mileposts, pullouts, guard walls, drainage culverts, survey points, wayside exhibits, avalanche chutes, trailheads, retaining walls, and intersecting roads. The park service linked this information to other databases that described the condition of the road and historic guard wall. Developing a GIS database with help from GPS software will provide a consistent map to help engineer and design repairs and identify cultural and historic items for preservation.

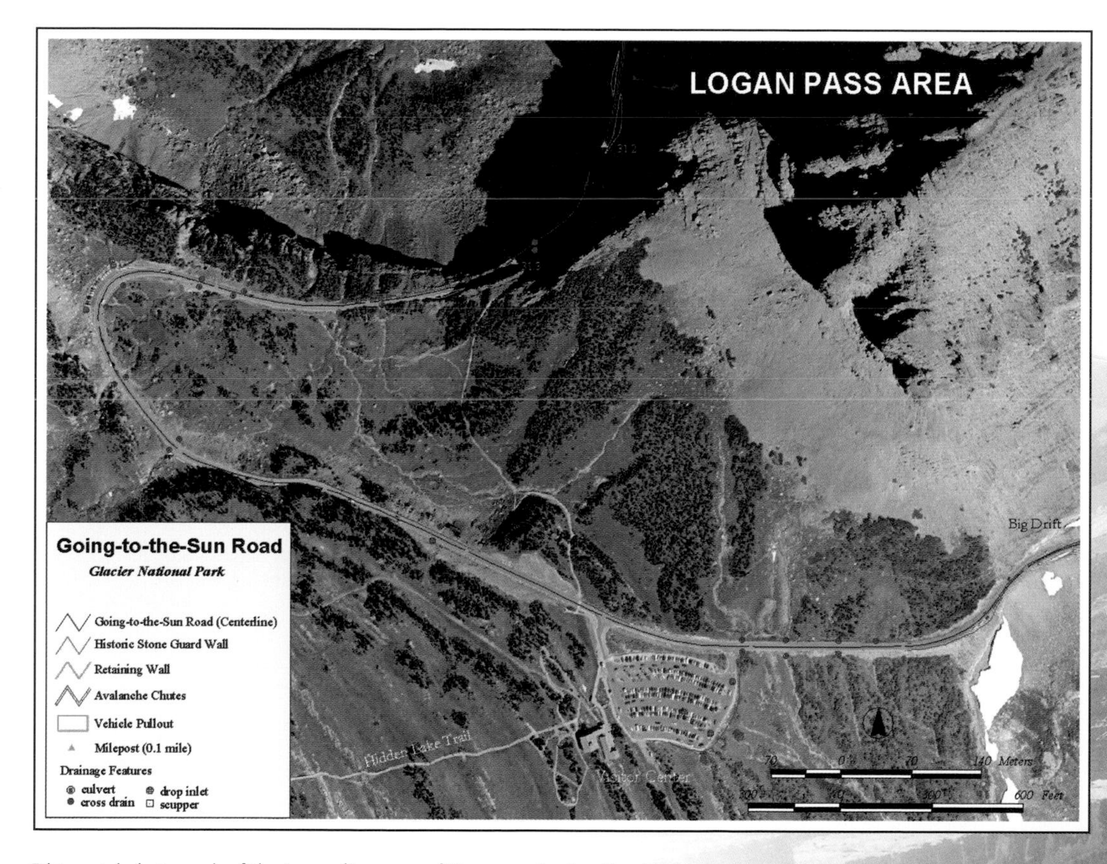

This aerial photograph of the Logan Pass area of Going-to-the-Sun Road is overlaid with information that illustrates road and drainage features. Logan Pass is the apex of Going-to-the-Sun Road and a key attraction for most park visitors. It demands a great deal of road-management attention because of the historic guard wall's deteriorating condition caused by avalanches and drainage issues. The image shows the severity of the terrain and the abundance of mapped drainage features, while also showing the capacity for visitor use.

Going-to-the-Sun Road winds through Glacier National Park.

Habitat Analysis and Road Relocation following Hurricane Opal

Evidence of Hurricane Opal's fury at Gulf Islands National Seashore.

Hurricane Opal demonstrated the power of nature when the storm hit Florida's Santa Rosa Island in 1995. Sustained winds of 125 miles per hour and a 15-foot storm surge heavily damaged buildings, destroyed roadways, eroded shoreline, razed sand dunes, and filled wetlands with sand in this part of Gulf Islands National Seashore. To help document damage and rebuild the park, the National Park Service used GIS software and global positioning system technology to create maps. GIS helped show how the island literally "rolled over" on itself in the storm, moving 200 feet closer to the mainland in certain areas, with the gulf beach receding by as much as 150 feet. The hurricane gave the park service the perfect opportunity to reevaluate the island's main road, which had blocked the natural growth and migration of the dunes, fragmented their habitat, and disrupted a second dune system. GIS mapping illustrated the problem and helped convince the Federal Highway Administration to relocate three miles of the seven-mile-long road. The park is also using GIS to help identify areas prone to storm flooding and damage, and island features to avoid during future reconstruction.

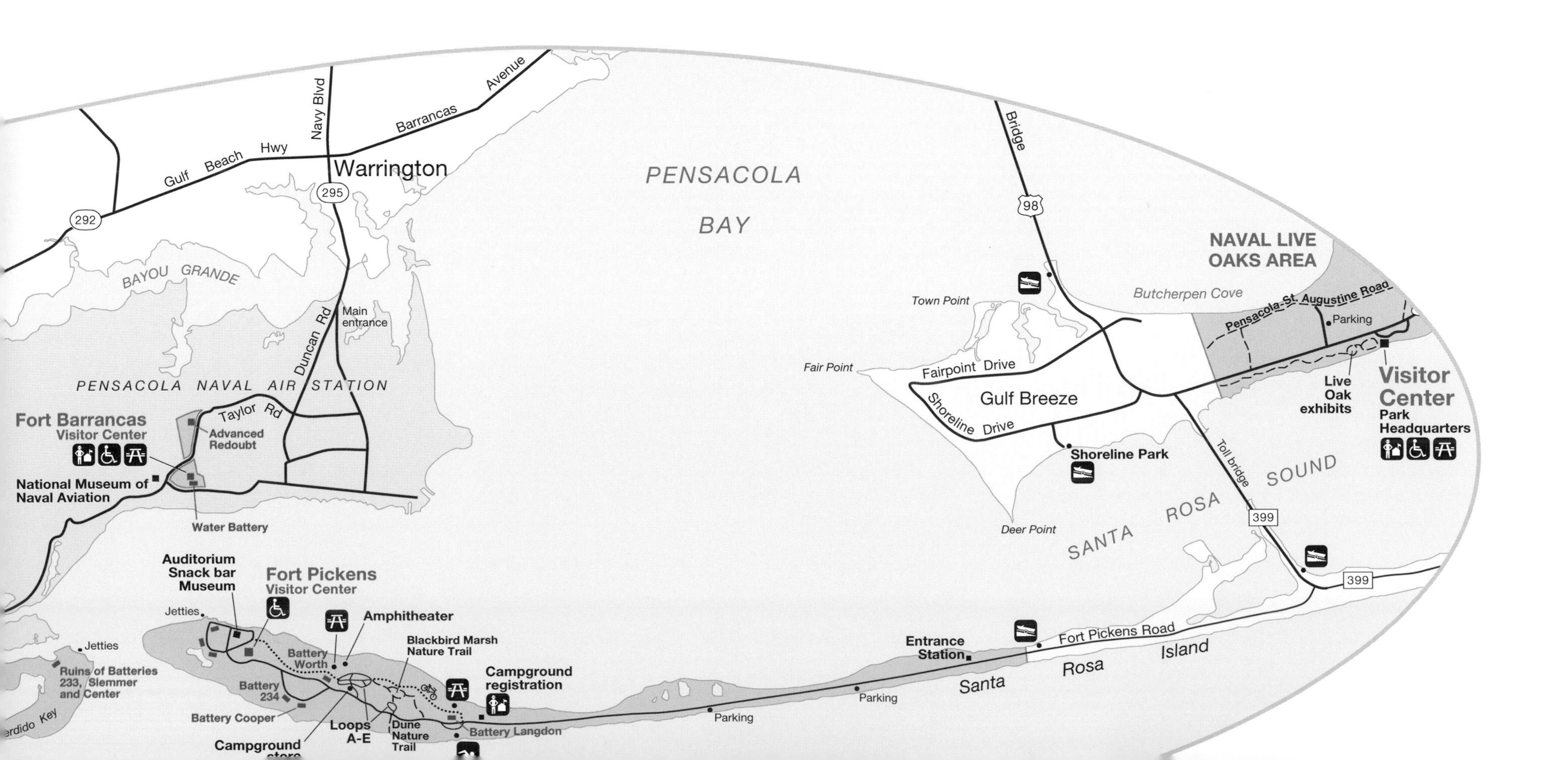

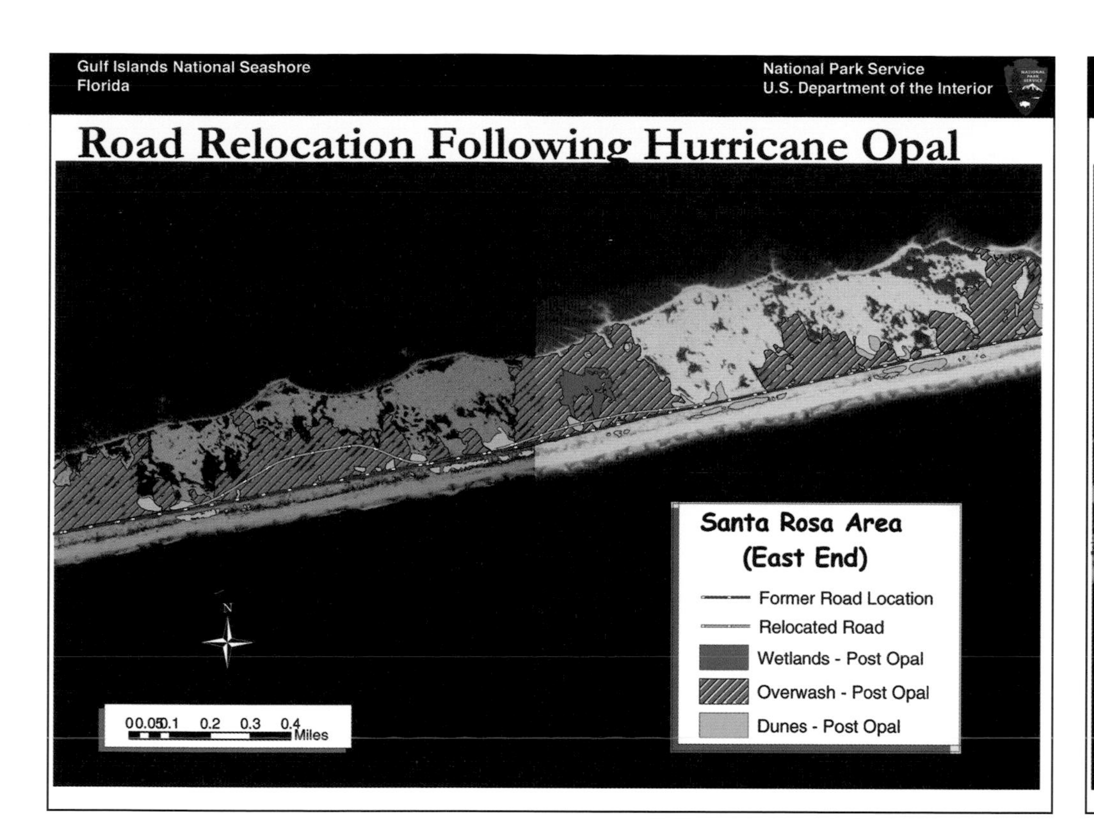

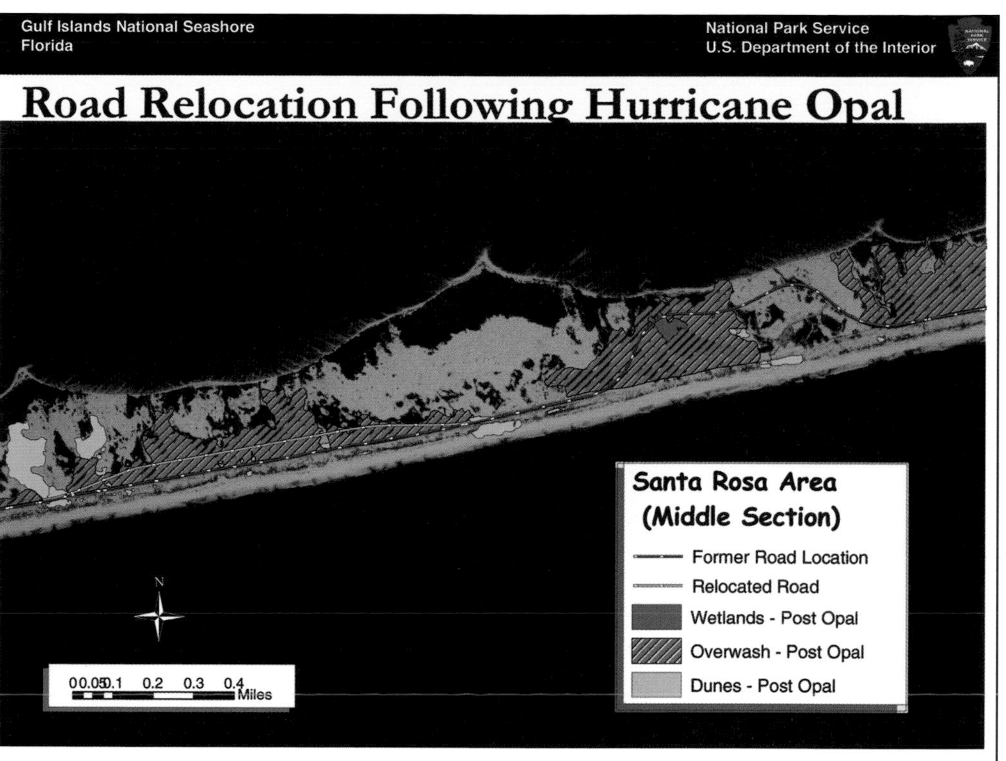

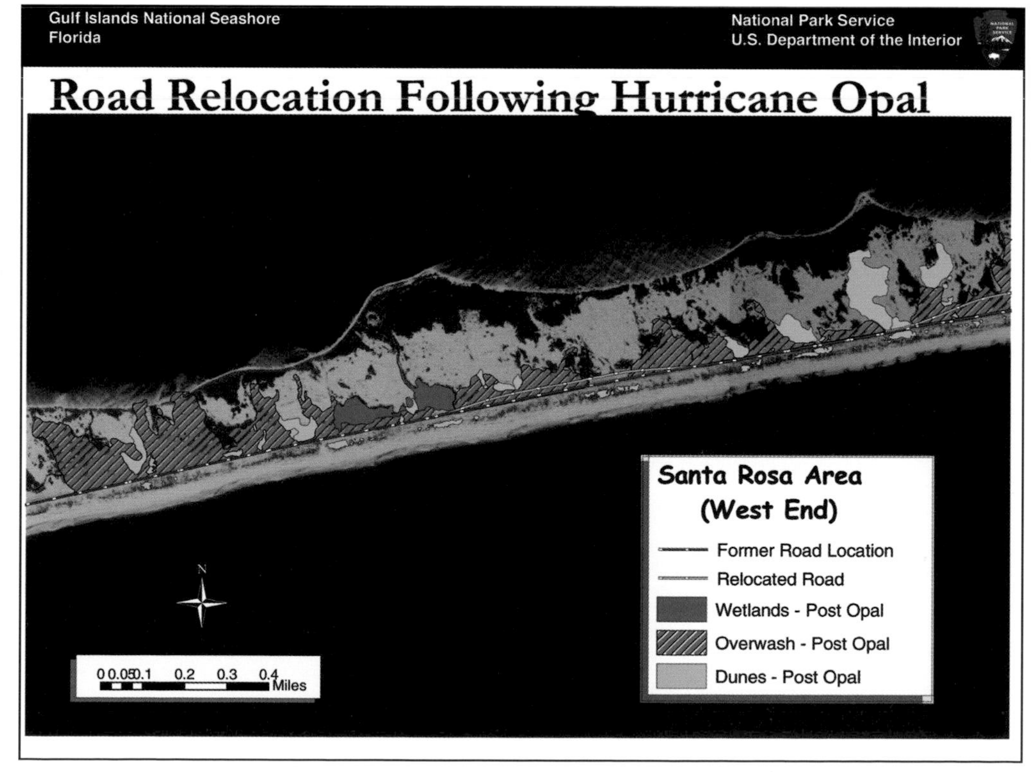

These storm impact maps show relocated road sections on the eastern, middle, and western parts of Santa Rosa Island.

Chapter 9

Leading the Way: Park Maps

This is the Information Age. . . . Parks are created to a higher standard; the National Park Service should be held to a higher standard. The Park Service must take whatever steps are required to secure the information needed to make sound management decisions.

Dr. Robin Winks, environmental scholar and former chair, National Park System advisory board

In their most basic form, maps show us where things are and how to get there. We pore over wrinkled road maps on summer vacations. We print out Internet directions to find the nearest post office or supermarket. At the mall, we check the directory to save our tired feet from a wrong turn. In national parks, maps point us to the geyser basin, natural bridge, or waterfall we came to experience. Maps and geographic information have always served another, larger purpose in the National Park Service: to preserve America's most special places for our benefit, enjoyment, and education.

In the early days, cartographers created hand drawings of parks to illustrate such features as wagon trails, cattle-grazing areas, wildfire perimeters, battlefield topography, and stream meanderings. Some of the earliest maps came from John Wesley Powell and his expedition through the Grand Canyon and from the U.S. Army Fifth Cavalry at Yosemite in 1896. Field maps from military cartographers and surveyors have helped national parks re-create authentic landscapes for visitors to see where their ancestors fought and died during the Revolutionary War, War of 1812, and Civil War.

Today, the park service combines the best of early cartography with the latest technology to collect and display geographic information. Relatively new mapping tools include GIS software, global positioning systems, satellite photography, and storage of national park maps on computers for easy access and review. The park service spends more than seventeen million dollars annually on geographic data. This grand endeavor of more than eleven thousand topographic map sheets represents more than eighty-four million acres. The park service and other partners together are mapping soils, plant life, geology, and other earth science themes at 388 national parks. The ongoing program eventually will produce the first standard set of digital national park basemaps.

Using GPS and handheld computers in the field, the park service can more accurately pinpoint locations of archaeological sites, endangered species, and nesting areas and then download them into GIS maps for review. A GIS map can show the geographic relationship of bird-nesting areas to soils, erosion rates, elevation, and vegetation. The ability to display all the information layers on a single map helps park managers as they decide where to release captive-bred California condors into the wild, close a trail to protect newborn bighorn sheep, or increase patrols against artifact theft.

Most mapping and geographic studies now extend beyond national park boundaries. They might display the relationship of the park to streams, coastlines, glaciers, rock formations, hazardous waste cleanup sites, and even cities. The park service works with federal agencies and other groups to collect geographic information using sophisticated sensors on satellites, aircraft, balloons, ships, and submarines. Displaying the information on GIS maps helps the park service compare different types of data. GIS technology can help park managers predict how various factors, such as drought, shoreline changes, or increased thermal activity, might affect a park. The power of GIS helps us relate, analyze, and model historic and current features and phenomena within a particular study area. Without new technology and geographic information, the National Park Service would face a difficult and sometimes impossible task to measure, map, and understand America's changing landscape and environment.

Data Access

This database provides students of Tlingit with map-based access to 201 recordings of Tlingit elders speaking the names of local places. Fewer than five hundred fluent speakers of the Native American Tlingit language survive. Students can click on a point, hear the name, view photos of the location, and read the stories that connect the people to the land. The photograph in this poster shows a sea arch at Ghaanaxhaa on the outer coast of Glacier Bay National Park and Preserve, an important cultural site for the Tlingit Takdeintaan clan. Ghaanaxhaa copyright the Hoonah Indian Association, used with permission. Sea arch photo courtesy of Wayne Howell, National Park Service.

Glacier Bay National Park and Preserve connects us to wild Alaska, now and for all time. Its tidewater glaciers, deep fjords, mossy rain forests, rugged coastline, and snow-capped peaks provide a living laboratory to explore the natural world. Its diverse landscapes and seascapes support wildlife ranging from humpback whales, porpoises, and harbor seals to black bears, wolves, and tufted puffins. A mosaic of plant life blankets the coastal and alpine regions, including alder, spruce, heath, spongy muskeg, and a few species in areas released from the grip of glaciers. These complex ecosystems present park managers with scientific information ranging from detailed undersea profiles to the distribution of bear habitat. The national park at Glacier Bay uses GIS technology to display complex and voluminous information in map-enabled databases for fast and easy access to data, aerial and ground photographs, sound files, and video. These databases provide instant spatial access to tens of thousands of photos, many taken in remote areas that the average park manager may never visit. A manager can zoom in on the map, click on a point, review all the photos taken there, see graphs of scientific data, view a species list specific to that site, even read the notes taken by the field crew at the site. One database stores information about the distribution of whales in Glacier Bay, helping managers decide where and when to restrict seagoing vessels in the summer tourist season. Another database of bear sightings and bear–human interactions helps managers decide when to close remote campsites to protect visitors and their belongings. Another project includes a talking map that helps Native American students learn the traditional names of local places and provides them a crucial link to the cultural landscapes of Glacier Bay. These are just some of the ways the national park uses databases supported by GIS technology.

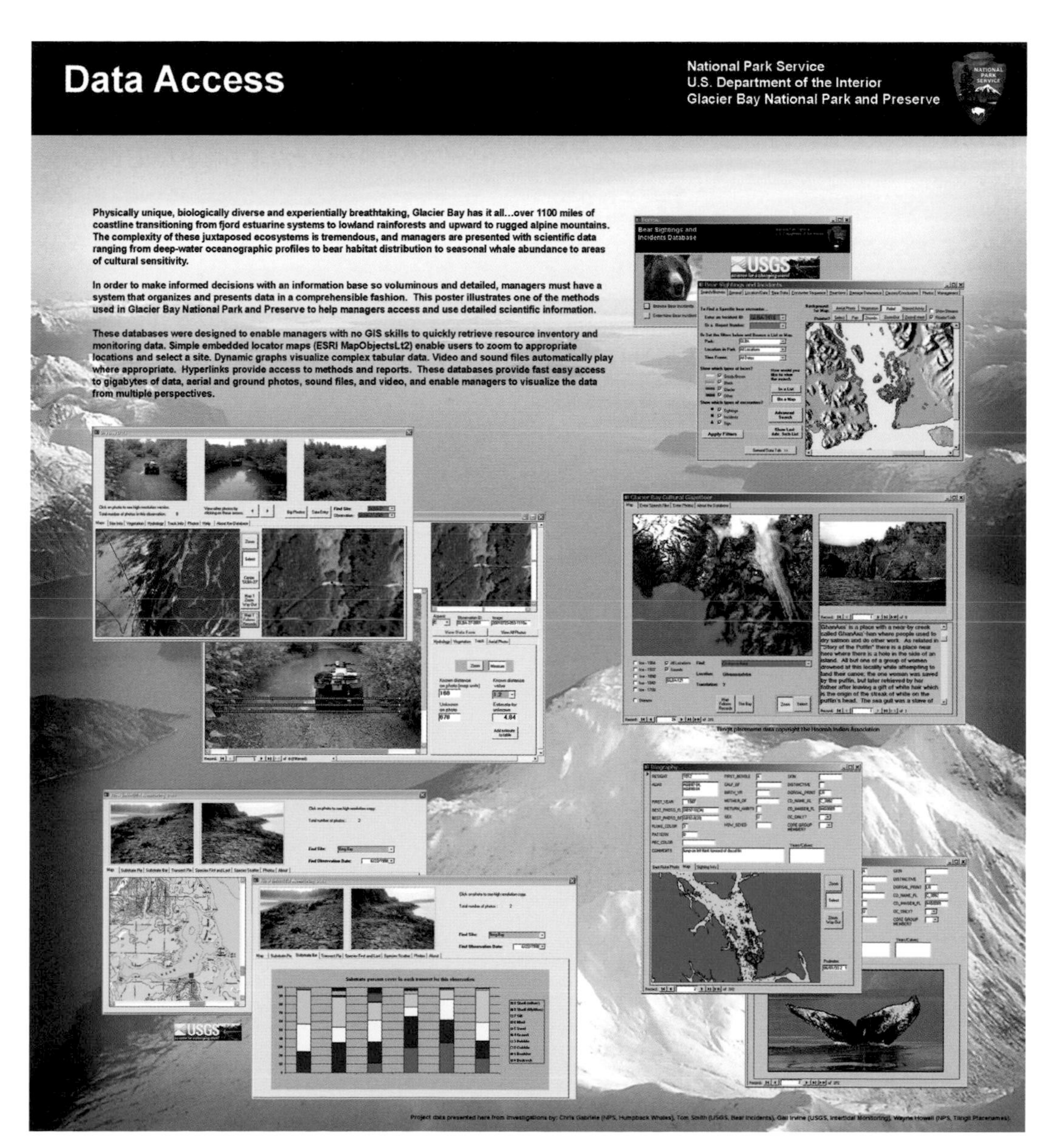

This poster created for the National Park Service 2003 Spatial Odyssey Conference shows databases used at Glacier Bay to store and display resource inventory and monitoring data. The poster demonstrated the integration of maps, photos, video, sound, graphs, and supporting documents.

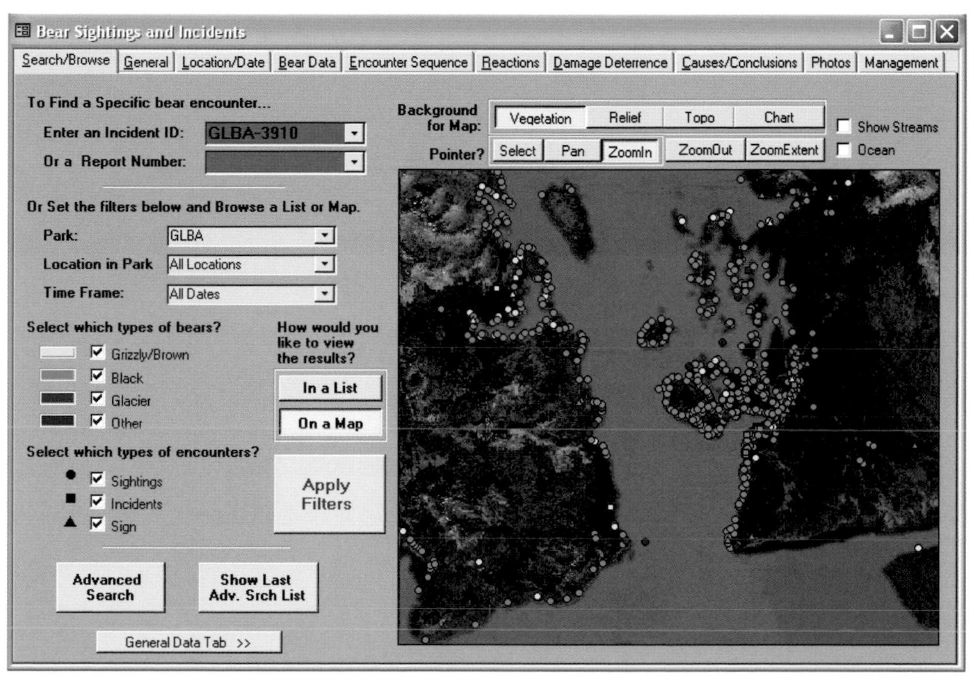

The map above shows all recorded bear sightings and incidents in lower Glacier Bay from 1952 to present. The clustered sightings on the eastern shore occurred next to the Bartlett Cove developed area near popular kayaking spots in the Beardslee Islands. Managers can browse to a specific location and then view the data relevant to that location. The background image in the map shows the vegetation.

Natural Heritage Occurrences, Colonial National Historical Park

CHARLES D. RAFKIND, NPS PHOTO

ANNE CHAZAL, VIRGINIA DEPARTMENT OF CONSERVATION AND RECREATION, DIVISION OF NATURAL HERITAGE

The National Park Service originally prepared the poster on the facing page for a session of the Northeast Coastal and Barrier Network. The meeting brought together scientists and natural resource specialists from the park service, other federal agencies, and research partners from academia. The poster presents an overview of the natural resources and issues of Colonial National Historical Park, including visitor use, endangered species, shoreline change, and damage. The positive response from park interpreters resulted in the adaptation of the poster for display in the park's visitor centers.

VIRGINIA DEPARTMENT OF CONSERVATION AND RECREATION, DIVISION OF NATURAL HERITAGE

Colonial National Historical Park in Virginia offers a variety of natural features, including marshes, seasonal ponds, streams, and swamps. Above, Virginia state zoologist Christopher Hobson surveys the park's biological resources, which include the rare skipper butterfly, shown at left. A scene of Powhatan Creek appears at top left.

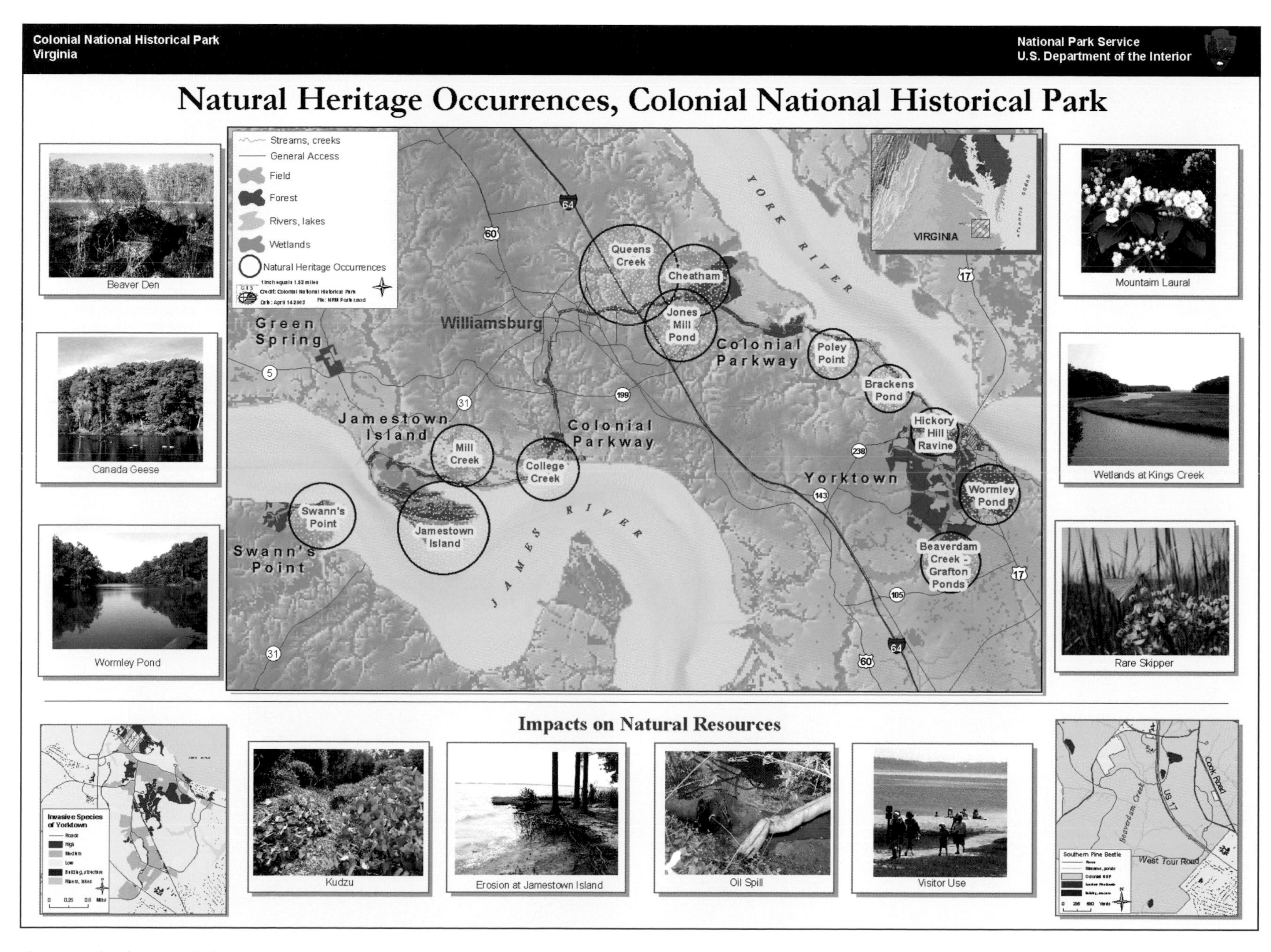

This poster identifies wetlands, forests, rivers, and other natural resources. Photographs show the natural beauty of the park, as well as the impact of oil spills, recreation, and other human activity.

Enhanced Wetlands Mapping at Point Reyes National Seashore

SUE VAN DER WALL, NPS PHOTO

This hierarchical classification system describes wetlands based on many factors. These include the source of water, such as marine, freshwater lake, river, estuary, or groundwater; the class or type of vegetation or substrate occurring on the site; and the water regime, or duration and nature of flooding or saturation. Also included in the classification are descriptions of any human-caused modifications of the environment that affect the duration or nature of water on the site. Wetlands maps use a classification code, or National Wetlands Inventory code, because there are more than two hundred different types of wetlands. The National Park Service uses these detailed wetlands maps partly because they identify sensitive areas to avoid while building trails and other facilities. The maps also help the park service monitor rare habitats where threatened and endangered species live, identify degraded areas for possible restoration, and show natural and human-caused changes in habitats and ecosystems under the care of the park service.

Point Reyes National Seashore in northern California offers haven to thousands of migratory birds and more than fifty animal species, including elephant seals, gray whales, and tule elk. Its ocean views, winding trails, and historic lighthouse offer inspiration and recreation. Its rare coastal wetlands feature salt, brackish, and freshwater marshes, meadows, and seasonal ponds. This national seashore aims to preserve, protect, and restore these invaluable wetlands in the face of hazardous material spills, failing septic systems, beef and dairy operations, construction, and historic neglect and abuse of the land. As a preservation tool, GIS technology helps the National Park Service accurately map the location, size, and type of wetlands as a way to evaluate how they work, detect environmental damage, and plan restoration. The national seashore also will use GIS data to model the predicted effects of climate change on the plants and animals in the wetlands. In one case, field crews identified a new population of endangered plant species in wetlands of the four-thousand-acre area that feeds Abbotts Lagoon. The park displayed the wetlands and botanical inventory on a detailed GIS map for research, planning, and public review.

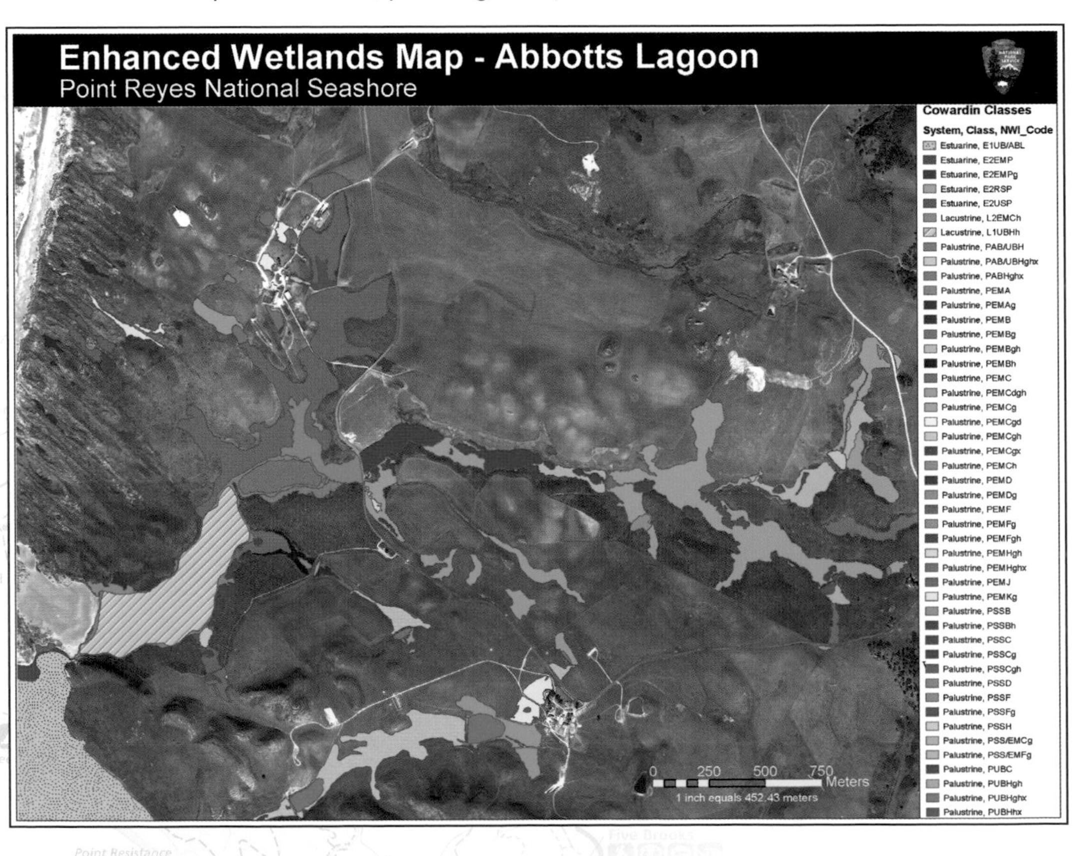

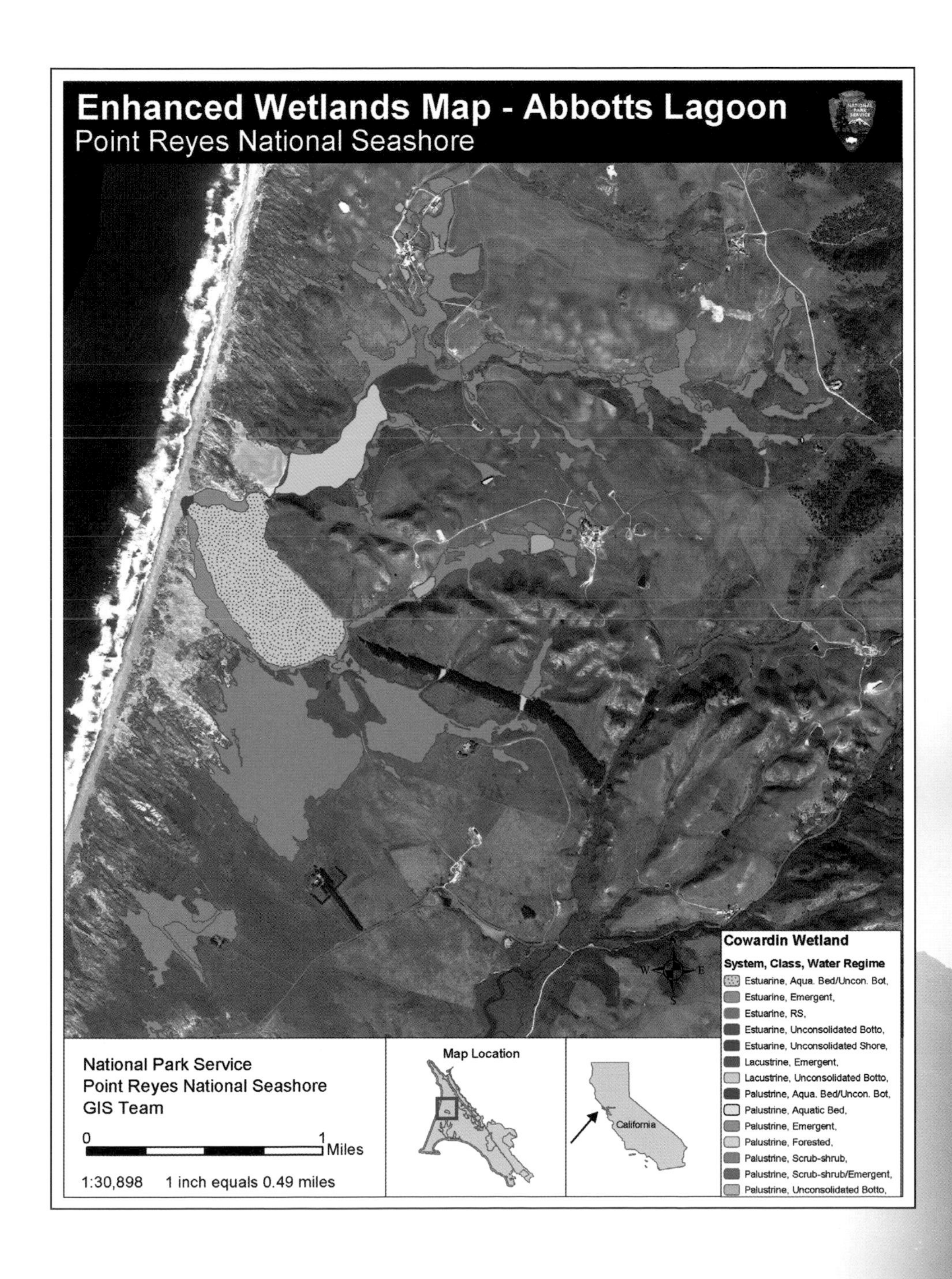

SUE VAN DER WALL, NPS PHOTO

White sandstone cliffs provide a dramatic backdrop to Drakes Beach at left. Built in 1870, the Point Reyes Lighthouse shown below warned mariners for 105 years before it was retired and transferred to the National Park Service for historic preservation (lighthouse photo courtesy of David Duran, National Park Service).

Flexible Image Classification with ArcGIS

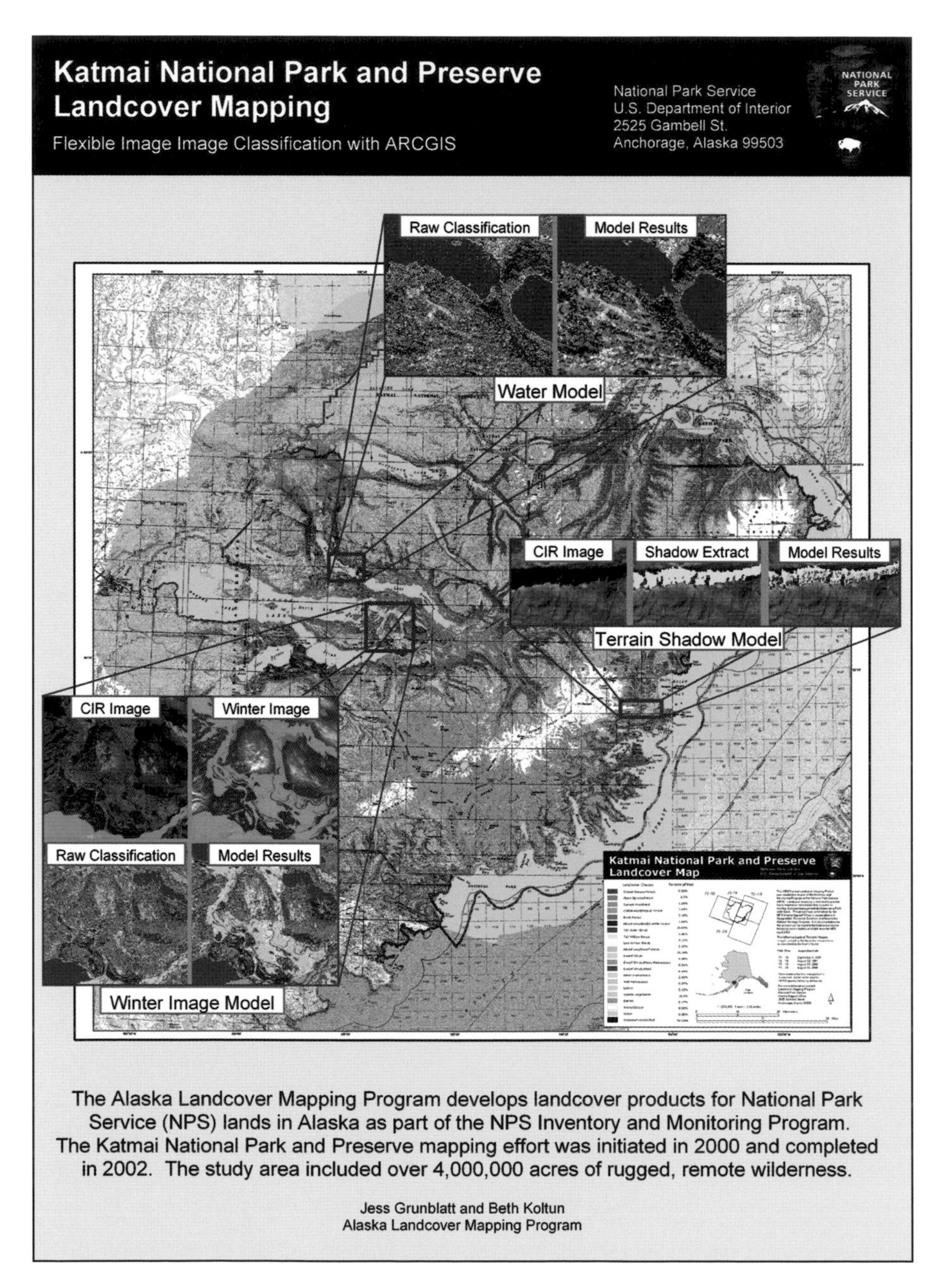

Katmai National Park and Preserve
Landcover Mapping
Flexible Image Image Classification with ARCGIS
National Park Service
U.S. Department of Interior
2525 Gambell St.
Anchorage, Alaska 99503
Raw Classification
Model Results
Water Model
CIR Image
Shadow Extract
Model Results
Terrain Shadow Model
CIR Image
Winter Image
Raw Classification
Model Results
Winter Image Model
Katmai National Park and Preserve
Landcover Map
The Alaska Landcover Mapping Program develops landcover products for National Park Service (NPS) lands in Alaska as part of the NPS Inventory and Monitoring Program. The Katmai National Park and Preserve mapping effort was initiated in 2000 and completed in 2002. The study area included over 4,000,000 acres of rugged, remote wilderness.
Jess Grunblatt and Beth Koltun
Alaska Landcover Mapping Program

This graphic illustrates the use of GIS modeling to enhance vegetation mapping within Katmai National Park and Preserve.

National parks within Alaska cover more than fifty-four million acres of some of the most pristine lands in America. From glaciers to grasslands and from polar bears to reindeer, these parks feature an incredible variety of habitat, wildlife, and plant species across vast landscapes. The National Park Service is using GIS and related technologies such as satellite imagery to develop land-cover maps for these large, rugged, and remote wilderness parks. Satellites record light reflected from the earth's surface as spectral data. Image-processing methods classify the spectral data into land-cover map classes. Interpretive GIS models further refine these land-cover map classes. These models allow computer-aided mapping to approximate some of the evaluations a photo interpreter would use during manual mapping. At Katmai National Park and Preserve, GIS models improved the identification of land-cover map classes and provided cost-effective mapping in a study area of more than four million acres. The final land-cover map provides reliable and consistent scientific information to evaluate the condition of ecosystems throughout the park and guide management decisions to ensure their preservation.

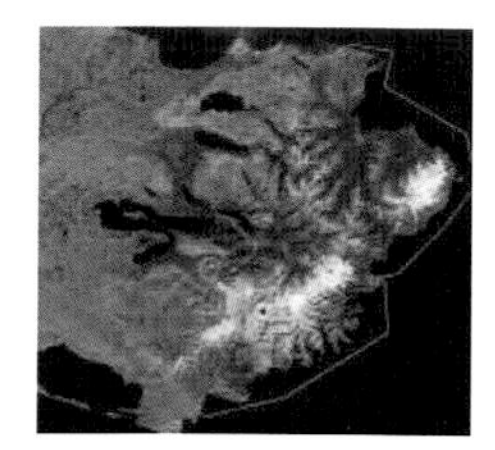
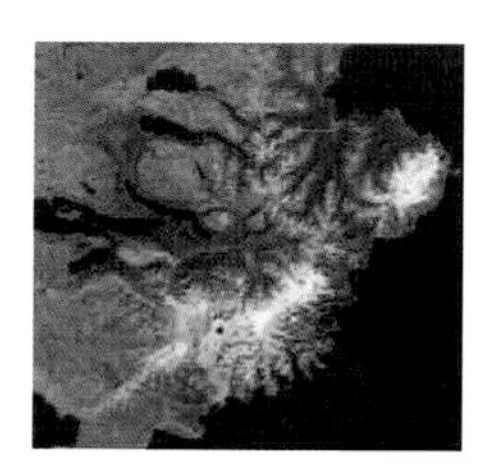
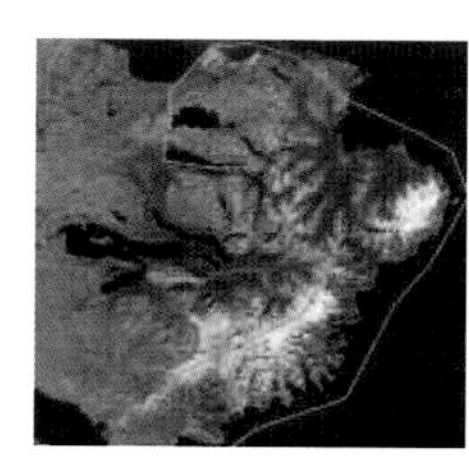

The abundance of vegetation species varies throughout the park. A series of regional masks applied during modeling incorporates this variation in the classification. Several of the masks are shown here using a color infrared mosaic of Katmai as background.

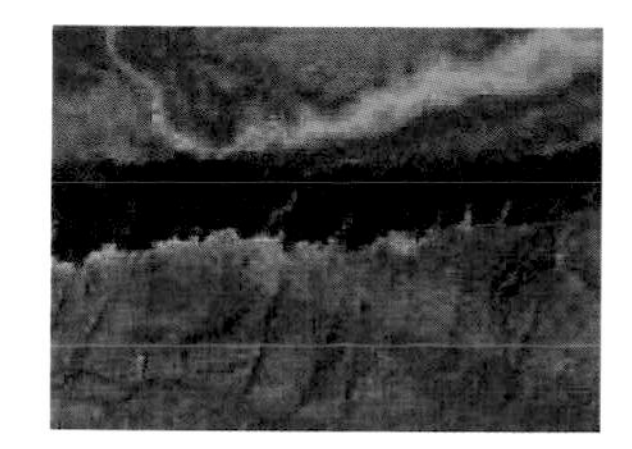

An area of shadow is shown for a color infrared representation of satellite data. After the preliminary classification, the areas of shadow within the study area were extracted and map classes within the shadow areas were further evaluated. Confusion classes were isolated and reclassified.

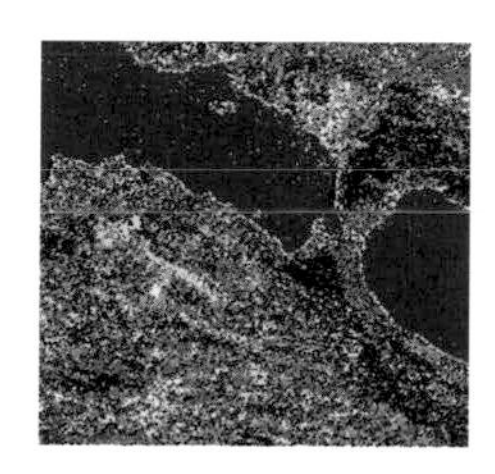
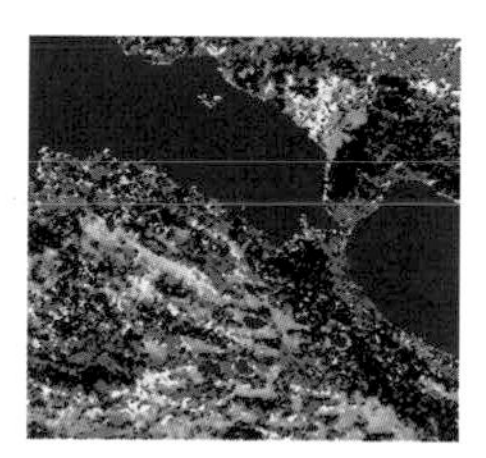

Water bodies were refined using a series of models. The original classification is presented at the left (water is blue), while the final land-cover map is presented at the right.

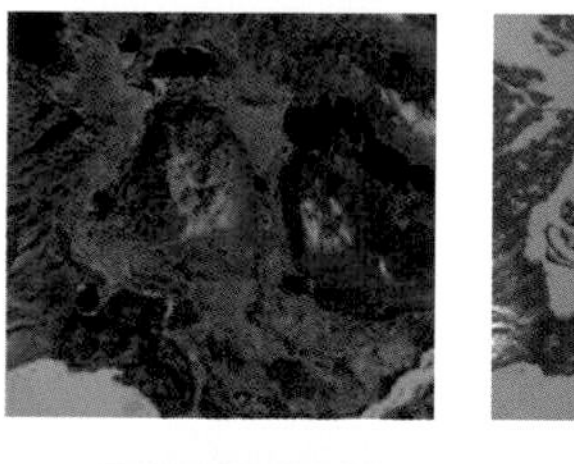
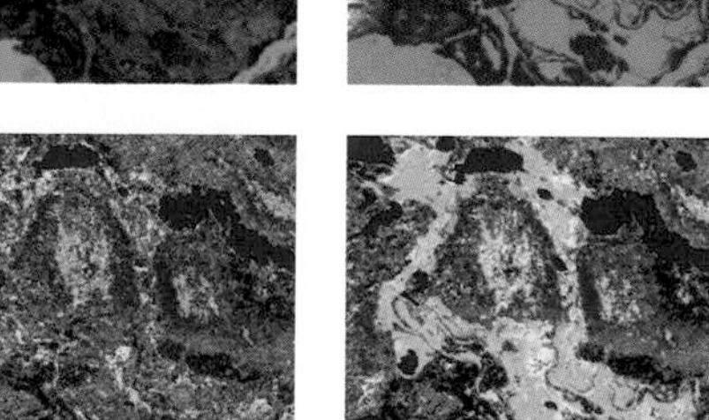

The upper-left picture depicts the Savonoski River area within Katmai that is characterized by large areas of wetlands. These wetlands are clearly seen as white in the upper-right picture from the winter scene of the same area. The original classification is shown at the lower left and much of the wetlands area is misclassified as conifer (dark green). Later modeling of wetlands using the winter image resulted in better delineation of these wetlands (light green).

Map Products

The unique GIS map "Existing Public Recreation Lands in Alaska and Yukon Territory" shows all the public lands of Alaska on one sheet. The map provides vital information for regional planning in the state's national parks, based on data collected from state and federal agencies and the Internet. It is one of a series of digital maps distributed to national parks in Alaska that helps scientists and managers use GIS technology in their daily work. Another

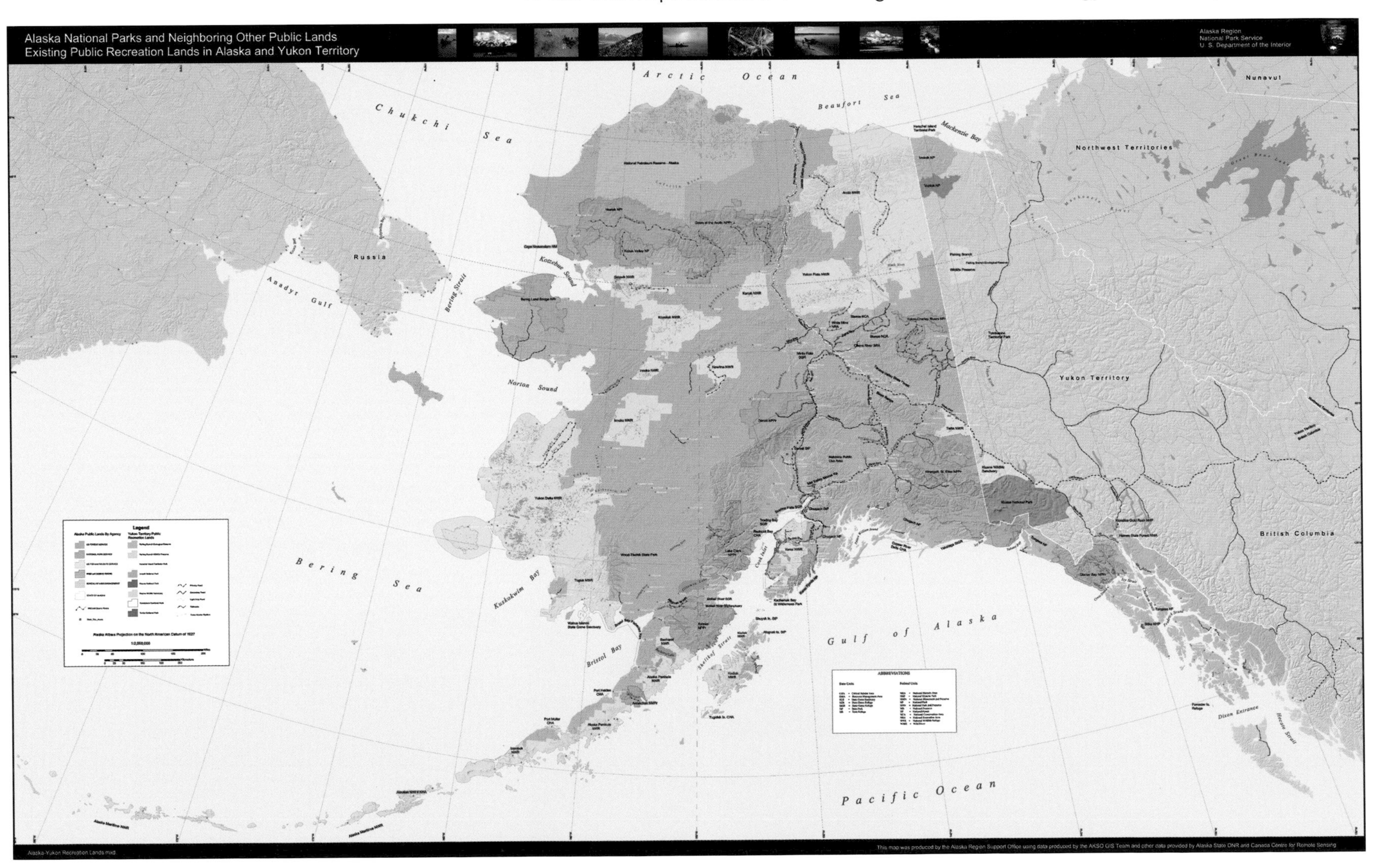

Alaska national parks and neighboring public recreation lands.

popular GIS map shows the national parks of Alaska in relation to federally designated Wild and Scenic Rivers. The park service also produces GIS maps on paper, including several statewide maps and a series of park maps for many uses, including wilderness and fire management.

National parklands in Alaska.

Facility Management in a Historic Settlement

This map identifies structures for possible historic preservation in the settlement at Kalaupapa National Historical Park.

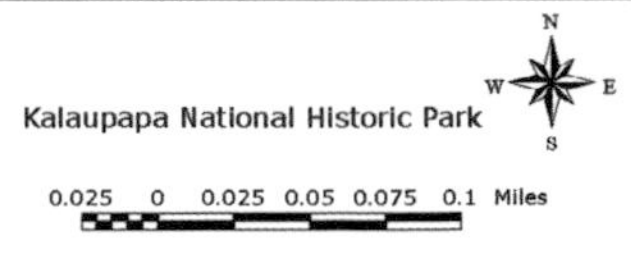

Established in 1980, Kalaupapa National Historical Park contains the setting for two tragedies in Hawaiian history. The first was the forced removal of indigenous Hawaiians in 1865 from Kalaupapa peninsula on the island of Molokai, where they had lived for more than nine hundred years. The second occurred between 1866 and 1969, when thousands of people diagnosed with leprosy across the Hawaiian Islands were forced to relocate to the remote peninsula. Many surviving patients of Hansen's disease, or leprosy, still live at the Kalaupapa settlement. Hundreds of historic buildings remain, and distinct neighborhoods support many activities of daily life. The National Park Service and state of Hawaii together protect the historic buildings in the settlement and preserve the cultural landscape of the neighborhoods. The park inspected the buildings and used information from national databases to determine the historic value of each structure and the cost to repair more than two hundred of them. GIS software displayed the location of each historic building on an interactive map. The map contains data about the condition of each building and the work and money needed for restoration. The ability to join GIS data to other databases has given managers a new tool to preserve history at Kalaupapa.

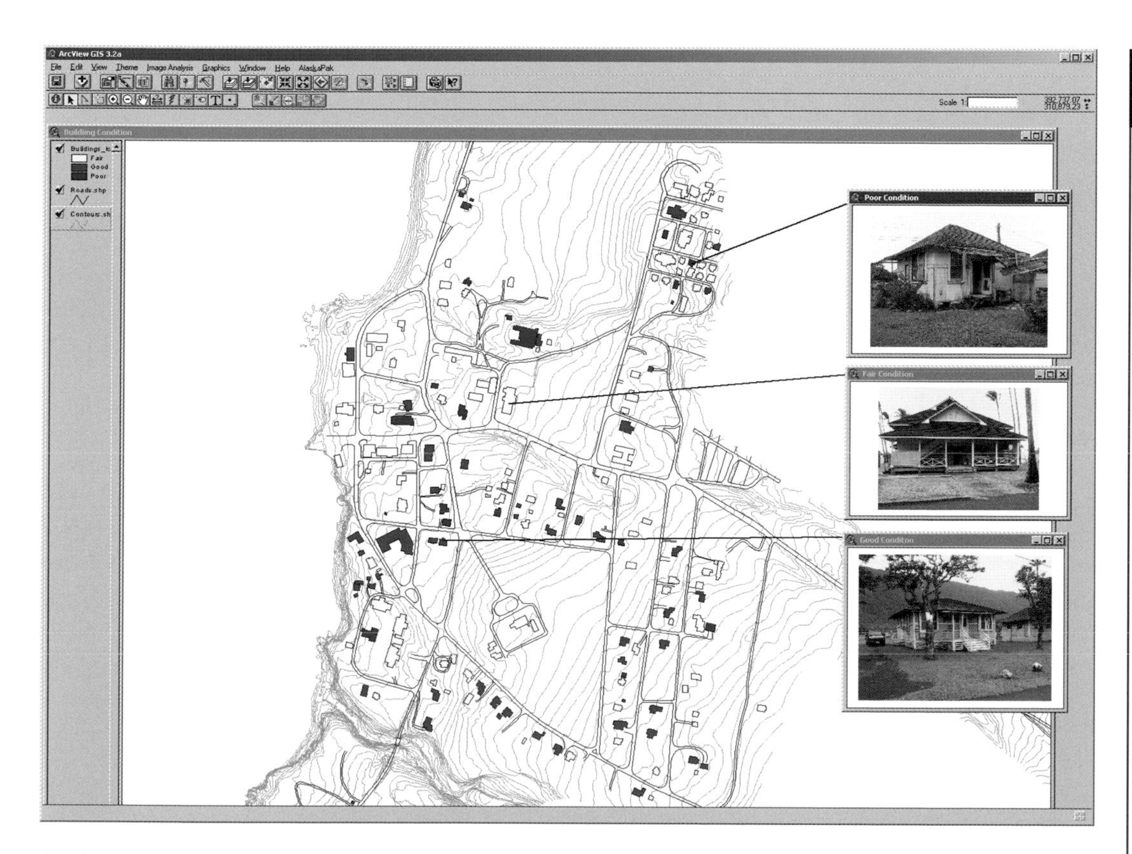

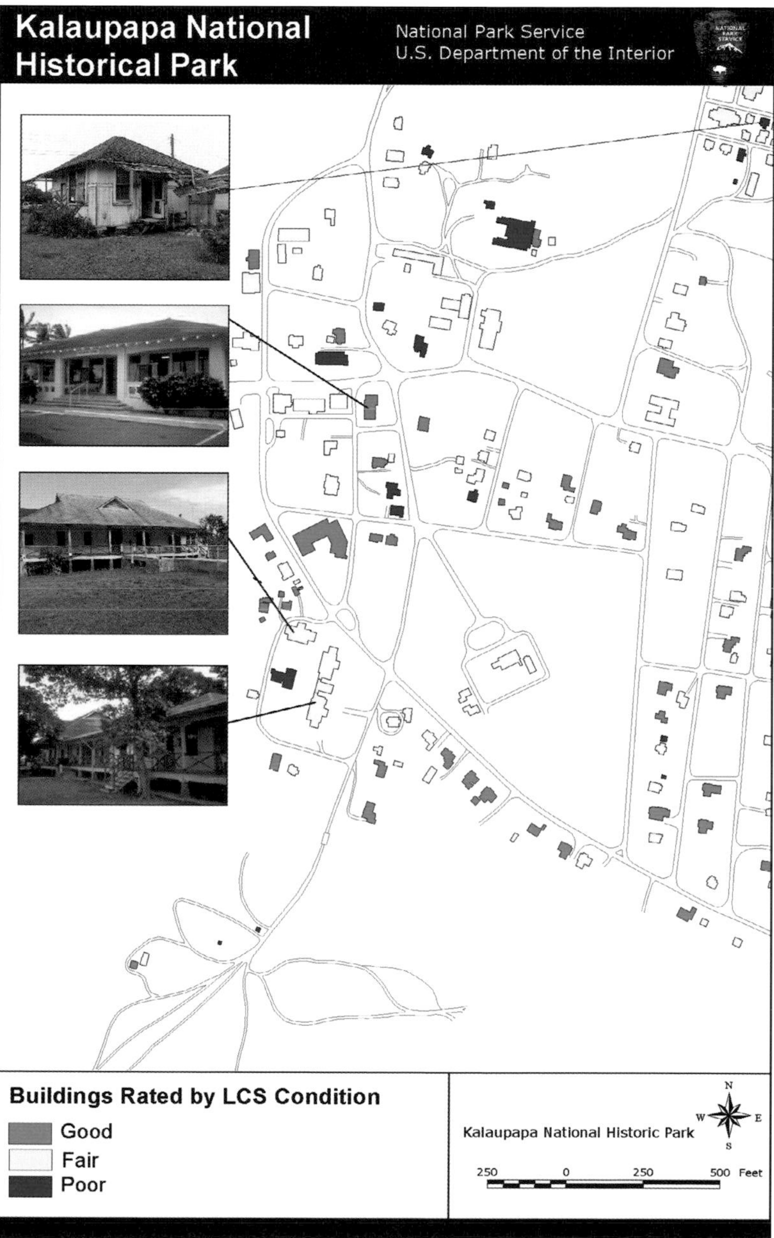

A color system identifies historic structures rated in good, fair, or poor condition in the national park. Photographs (above and at the right) show examples of each condition. The displays help park managers make decisions about the preservation of the Kalaupapa settlement.

Geology . . . It's Not Just for Scenery Anymore

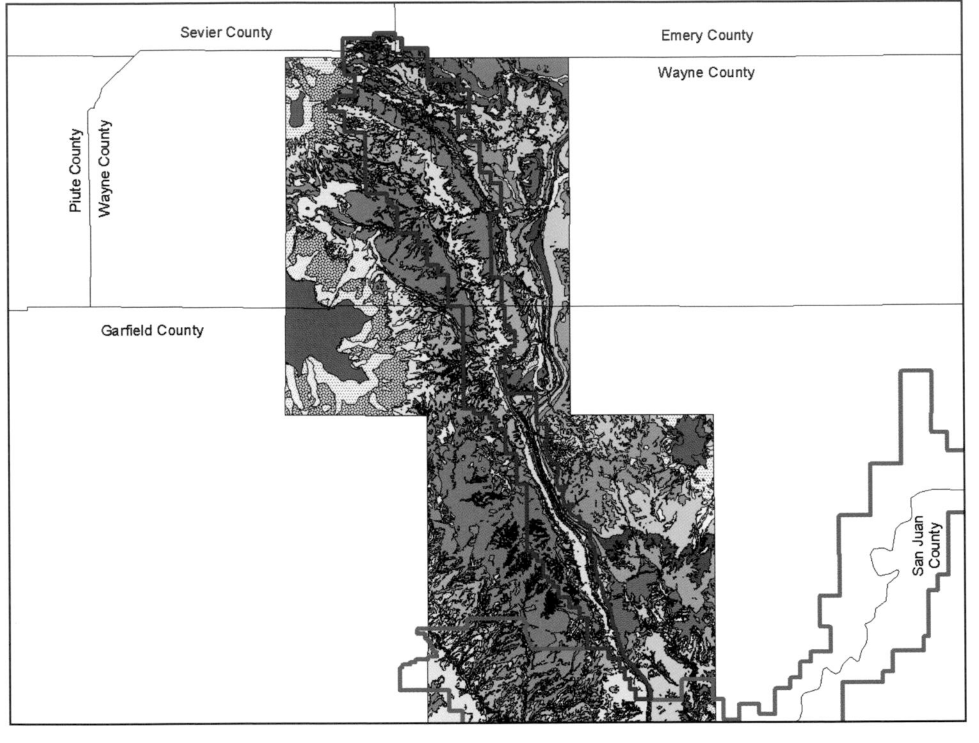

Digital geologic maps like this one of Capitol Reef National Park have several advantages over paper geologic maps. The park service integrates digital maps with other geospatial data (soils, vegetation, hydrology, and so on) to analyze spatial relationships. A digitally generated GIS map provides quick, reproducible, and precise analysis. Digital geologic maps also make it easier for users to share and transfer data. Park boundaries are shown here in red.

GIS and cartographic tools are particularly important to the geologists and other natural resource specialists working in national parks. Geology, often scoffed at as so much "scenery," actually plays a much larger role in the ecosystem. The importance of geology to our everyday activities is obvious to the geologist: it ranges from shaping the earth's surface to controlling where we eventually settle in communities. With GIS, the geologist can illustrate the importance of the science of geology to others, including soil scientists, botanists, and ecologists, as well as national park visitors. At Capitol Reef National Park in Utah, geologic GIS maps help predict habitat for threatened and endangered species. They include Winkler cactus, a protected species associated with the Morrison Formation of rocks dating to the Jurassic period; Barneby reed-mustard, associated with the Moenkopi Formation from the earlier Triassic period; and Jones cycladenia, associated with the Chinle Formation from the Triassic period. The park also found a direct link between the presence of several threatened and endangered plant species and outcrops of Navajo sandstone from the Jurassic period, when dinosaurs reached their maximum size. Geologic maps describe underlying conditions of natural systems and are key in the study of ecosystems, earth history, soils, and environmental hazards such as fires, landslides, and falling rocks. When combined with GIS technology, these maps have advantages over traditional paper geologic maps. GIS maps display the relationship between the earth's surface and bedrock and add to our understanding of soil, vegetation, water, and other features. GIS allows study of all these features on a single precise map for fast, easy review and creates a powerful database.

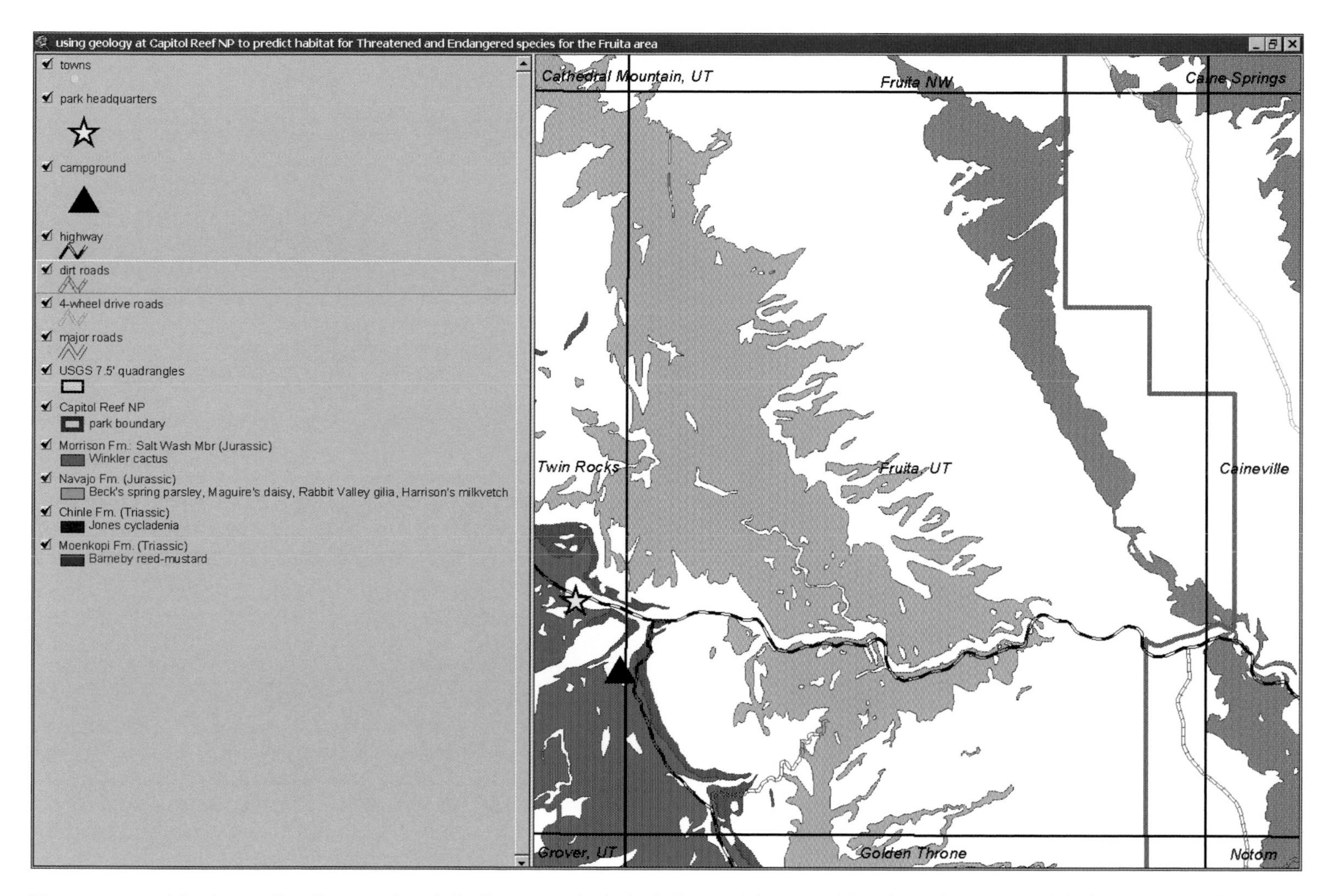

The occurrence and distribution of Winkler cactus (green), Beck's spring parsley (yellow), Jones cycladenia (purple), and Barneby reed-mustard (red), are shown in relation to geologic units and infrastructure in the Fruita area of Capitol Reef National Park.

JOHN BURGHARDT, NPS PHOTO

Head frame of the park's remote Duchess Mine, one of many abandoned uranium mines in Utah.

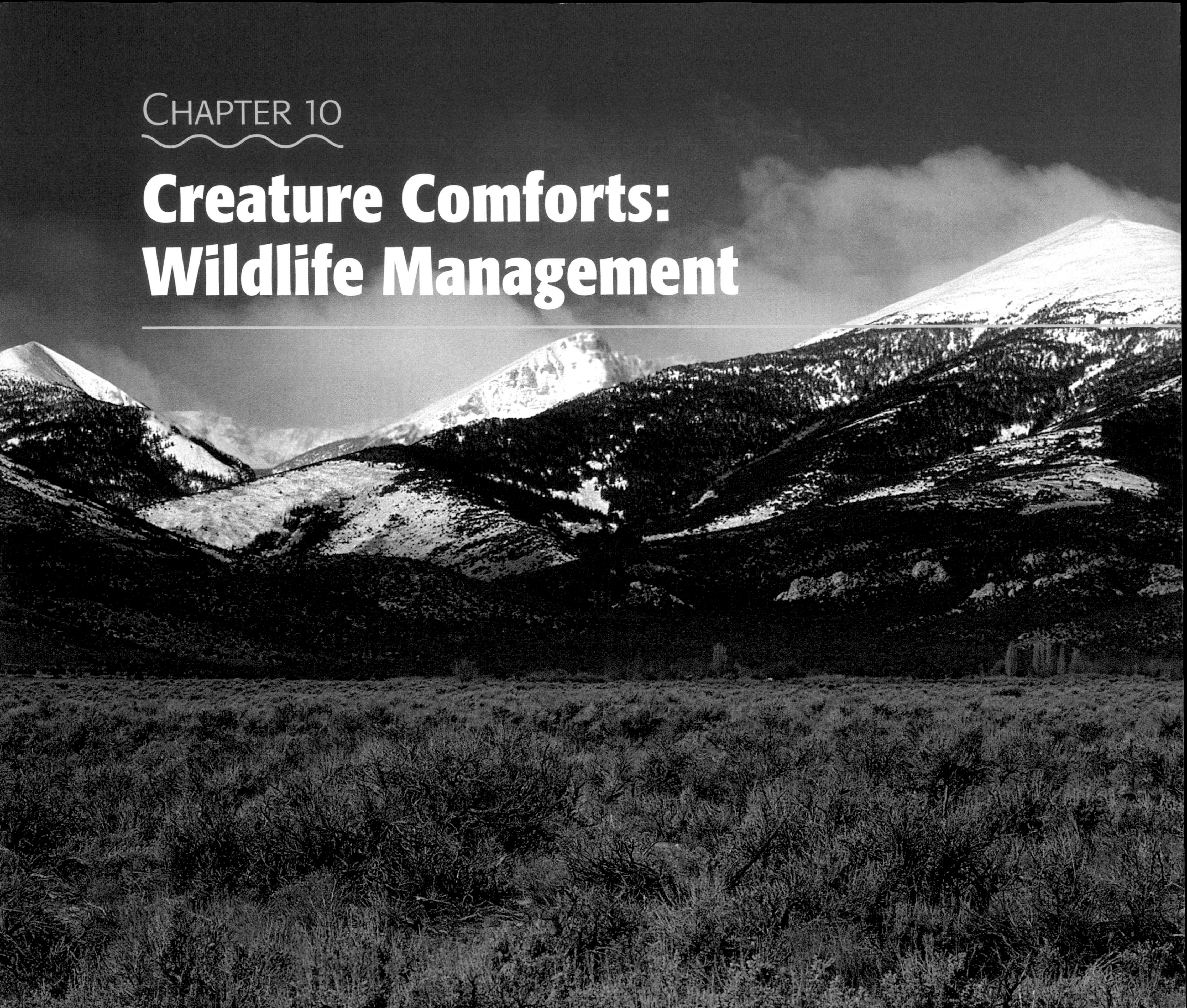

CHAPTER 10

Creature Comforts: Wildlife Management

But our national heritage is richer than just scenic features; the realization is coming that perhaps our greatest national heritage is nature itself, with all its complexity and its abundance of life, which, when combined with great scenic beauty as it is in the national parks, becomes of unlimited value. This is what we would attain in the national parks.

George M. Wright, Joseph S. Dixon, and Ben H. Thompson, *Fauna of the National Parks of the United States*, 1933

For the National Park Service, the future of thousands of plant and animal species depends on the preservation and restoration of wild places in our country. Habitat supports the California condor as it swoops over the Grand Canyon. It offers refuge to the desert tortoise that rests in its den at Joshua Tree National Park. And it quenches the thirsty roots of the prairie fringe orchid at Pipestone National Monument in Minnesota. In all, the NPS manages nearly eighty-four million acres of public land for thousands of plant and animal species. Of those, hundreds are threatened or endangered by declining populations, imperiled habitat, or both. In some cases, a species might survive only in a particular national park, making the protection of its homeland paramount.

As the National Park Service manages our parklands, it must answer a series of basic but crucial questions about the geographic relationships of species to their habitats. These spatial questions start with:

Where is the species located?

What is the extent of its habitat?

What environmental factors affect that habitat?

Each of these questions will include a series of more refined issues. This is the realm of GIS. Generations of biologists have used hand-drawn habitat maps, aerial photos, and topographic maps to prepare their analyses. Now they are making the transition to GIS to improve the quality and timeliness of their work.

As we apply GIS technology to national parks land management, we are reminded that managing habitat for endangered species poses unique challenges. For example, biologists might not reveal certain data to the general public, particularly if doing so would make a species vulnerable to poaching or harassment. Yet coworkers, researchers, and other agencies might have a legitimate need for otherwise secure information. During a wildfire, for example, firefighters can protect a species if they know where it lives. Today, the use of GIS to manage habitat reflects a growing trend in the park service as staff at each park decide on the best approach for each species.

Determining Foraging and Roosting Areas for Mastiff Bats (*Eumops* spp.) Using Radio Telemetry

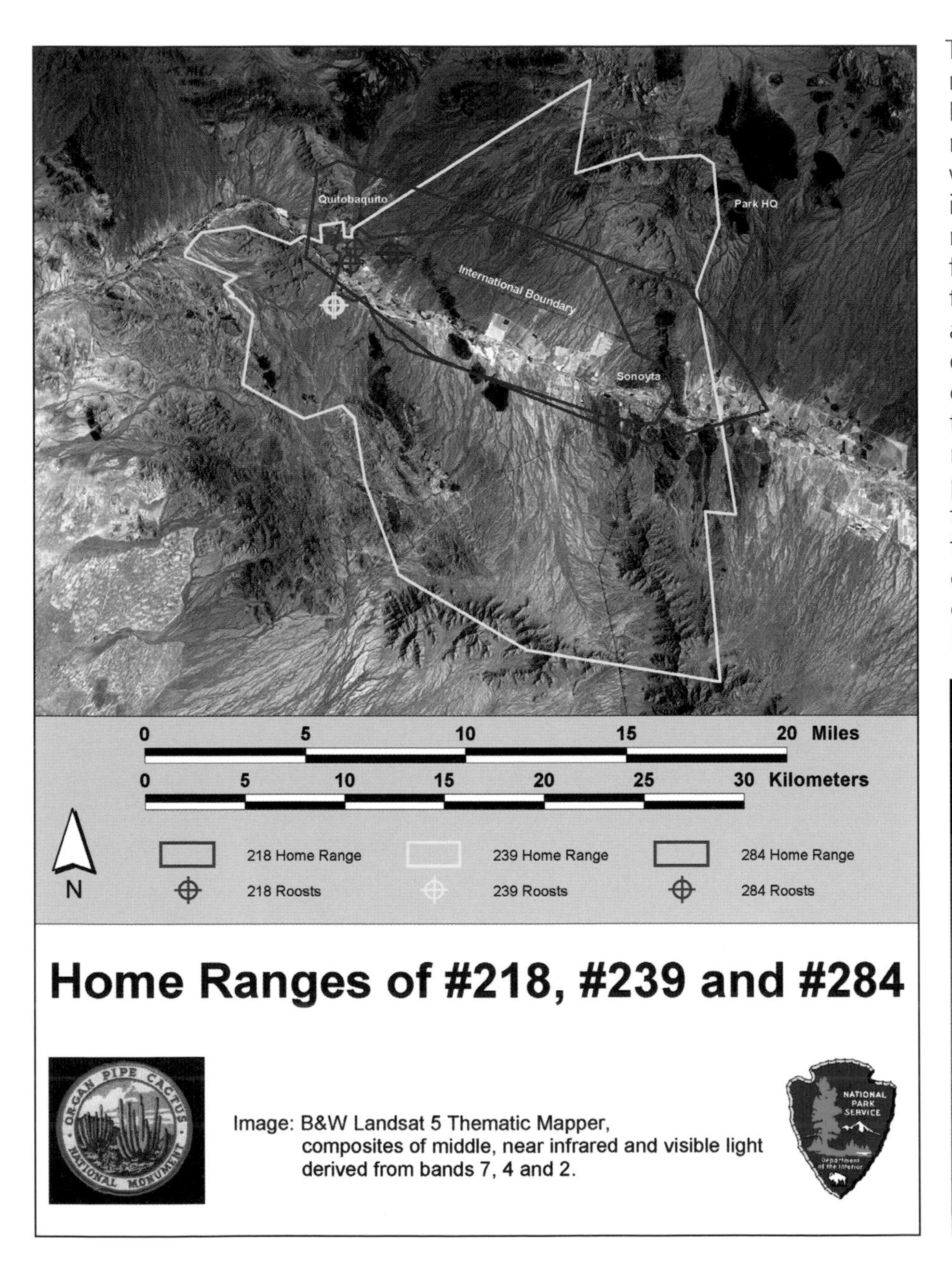

Using radio telemetry and signal triangulation, the park service and its partners identified the home ranges of three mastiff bats, tracking them from the moment they left a roost until they returned after hours of foraging in the Sonoran Desert.

Tracking bats as they zigzag through the night sky became the focus of a study to learn the roosting patterns, range, and routes of a little-understood tropical species found in Organ Pipe Cactus National Monument in southern Arizona. Wildlife researchers wanted to know more about the history and habits of Underwood's mastiff bat along the Mexican border, where the nocturnal creature faces increased threats from human population, tourism, industrialization, and other changes in land use. Using radio telemetry to track the bats, U.S. researchers in a cooperative effort with the adjacent Pinacate Biosphere Reserve in Mexico found that the species forages over wide areas of the Sonoran Desert along the international border, averaging thirty-seven square miles from rural to agricultural to semiurban areas on a typical night. Observers on ridgetops commonly heard the distinct vocalizations of nonradioed mastiffs overhead, sometimes seen and heard in groups of six or more, indicating that these bats fairly commonly forage along ridgelines, hilltops, and other raised topographic features. And for the first time, researchers learned from bats outfitted with radio transmitters that this species roosted in cavities carved out of saguaro cactus by woodpeckers. To illustrate the habits of the bat, researchers put their data on GIS maps that show the home ranges, the roosting areas, and the typical routes used by the three bats tracked for the study. The National Park Service and other agencies will use the maps to help them develop and improve long-term wildlife strategies for the species. They also created an educational poster of their findings and are translating it to Spanish for outreach at the sister Pinacate reserve and in schools across the border.

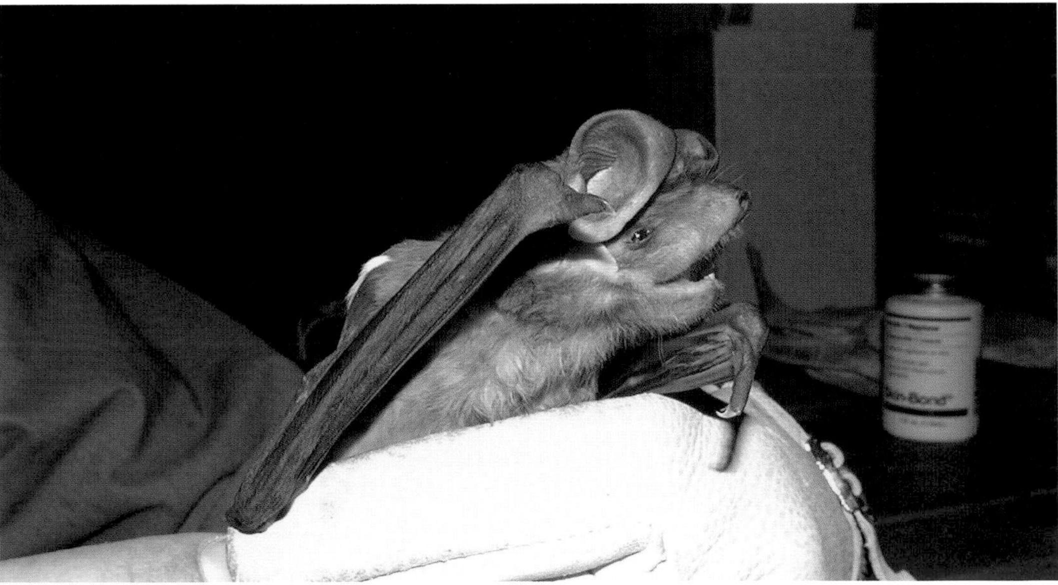

Researchers attach a radio transmitter to a mastiff bat at Quitobaquito Pond. Researchers then track the radio signal from atop hills along the U.S.–Mexican border.

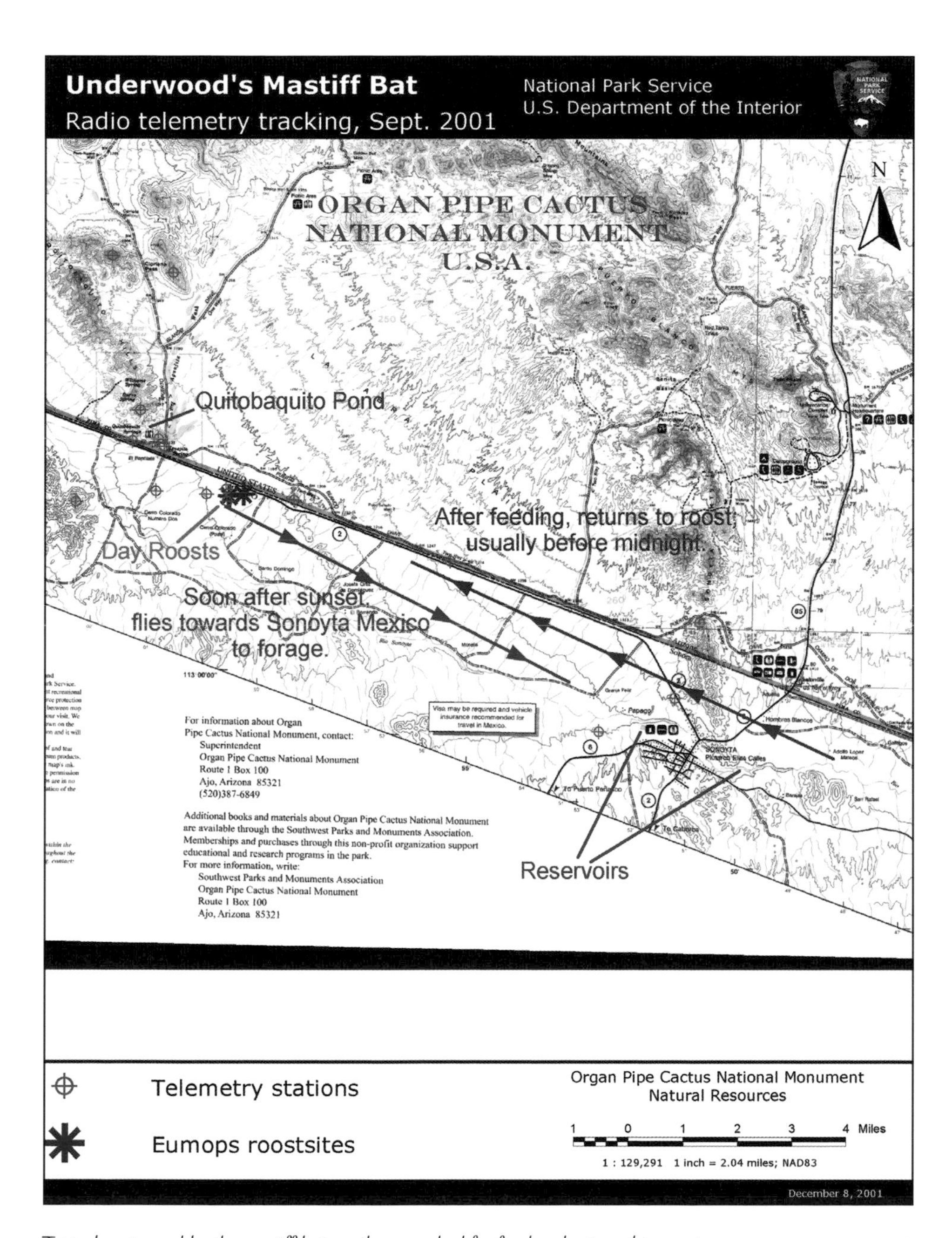

Typical route used by the mastiff bats as they searched for food and returned to roost.

Mexican Biosphere researcher Juan Miranda locates a bat roost using radio telemetry. This mature saguaro has dozens of cavities used by a mastiff bat and several bird species.

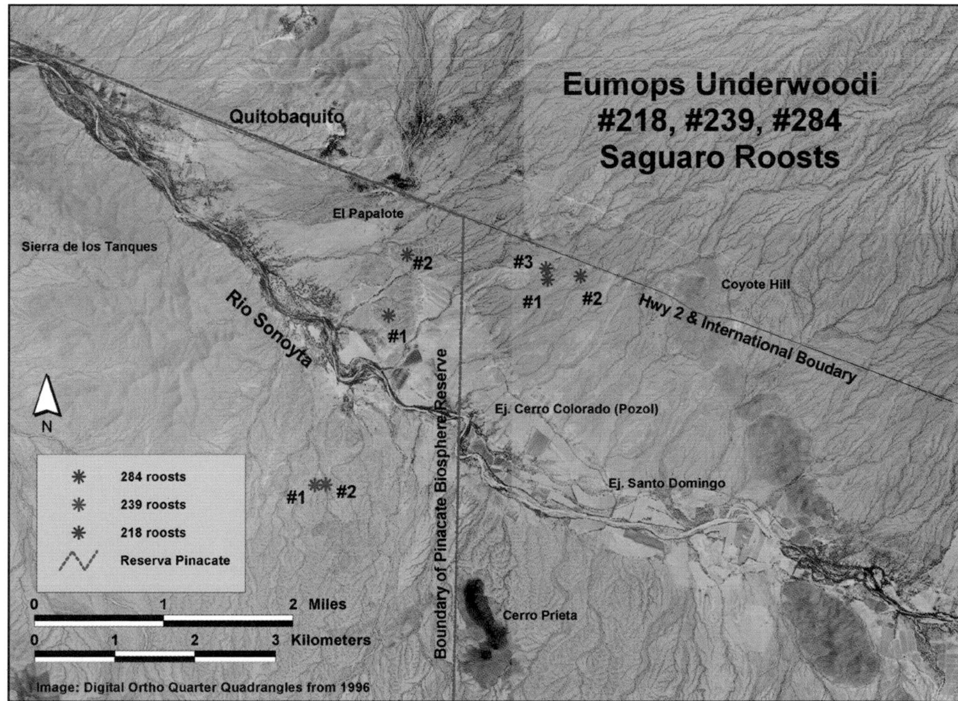

Radio telemetry helped identify roosting areas. Each bat used one saguaro cavity as shelter during the day, sometimes changing cavities after foraging at night. Quitobaquito Pond is the closest dependable large source of water.

Satellite Tracking of Endangered Kemp's Ridley Sea Turtles

Researchers attached satellite transmitters like the one shown here to track the movements of twenty-three Kemp's ridley sea turtles that nested in or near Padre Island National Seashore in Texas.

Tracking sea turtles from space has helped researchers predict when and where these endangered creatures will lay clutches of eggs off the coast of Texas in the Gulf of Mexico. The study used satellite telemetry and GIS mapping to follow twenty-three Kemp's ridley sea turtles that nested at or near Padre Island National Seashore, where wildlife biologists have spent more than twenty-five years trying to reestablish a turtle nesting colony. Nesting has increased in recent years, and more than half of Kemp's ridley nests located in the nation are found at the national seashore. Between 1997 and 2003, researchers outfitted twenty-three turtles with radio transmitters and then monitored their travels during and after the nesting season. In the nation's first study of its kind for this endangered species, the U.S. Geological Survey, National Park Service, and other research partners wanted to see how the turtles used their habitat and where they would lay successive egg clutches. The study also evaluated potential threats to the turtles in the national seashore and gulf. Researchers used data gathered in the study to create GIS maps that displayed turtle movements for easy review. After the nesting season, most of the turtles tracked in the study left south Texas and traveled north along the coast, with their last identified locations in the northern or eastern waters of the Gulf. The data helped predict when and where four turtles would lay their clutches, thus improving protection of the nesting turtles and their eggs. To protect sea turtles, Texas wildlife officials recently used the data to require a seasonal closure of south Texas Gulf waters to shrimp trawling. The information also helped researchers evaluate experimental efforts to create a second nesting colony.

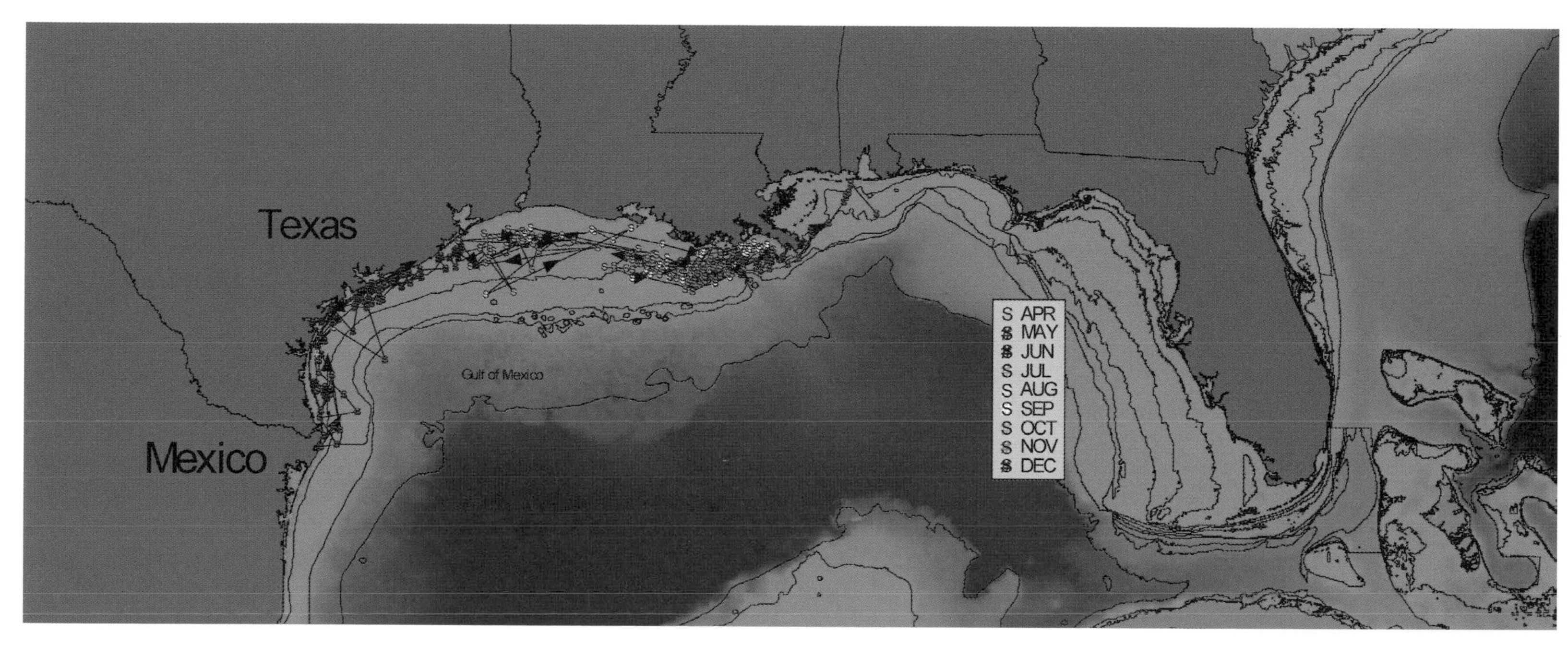

Satellite telemetry and GIS technology helped track Kemp's ridley sea turtles during and after nesting seasons in the Gulf of Mexico.

Effects of Urbanization and Habitat Fragmentation on Bobcats and Coyotes in Southern California

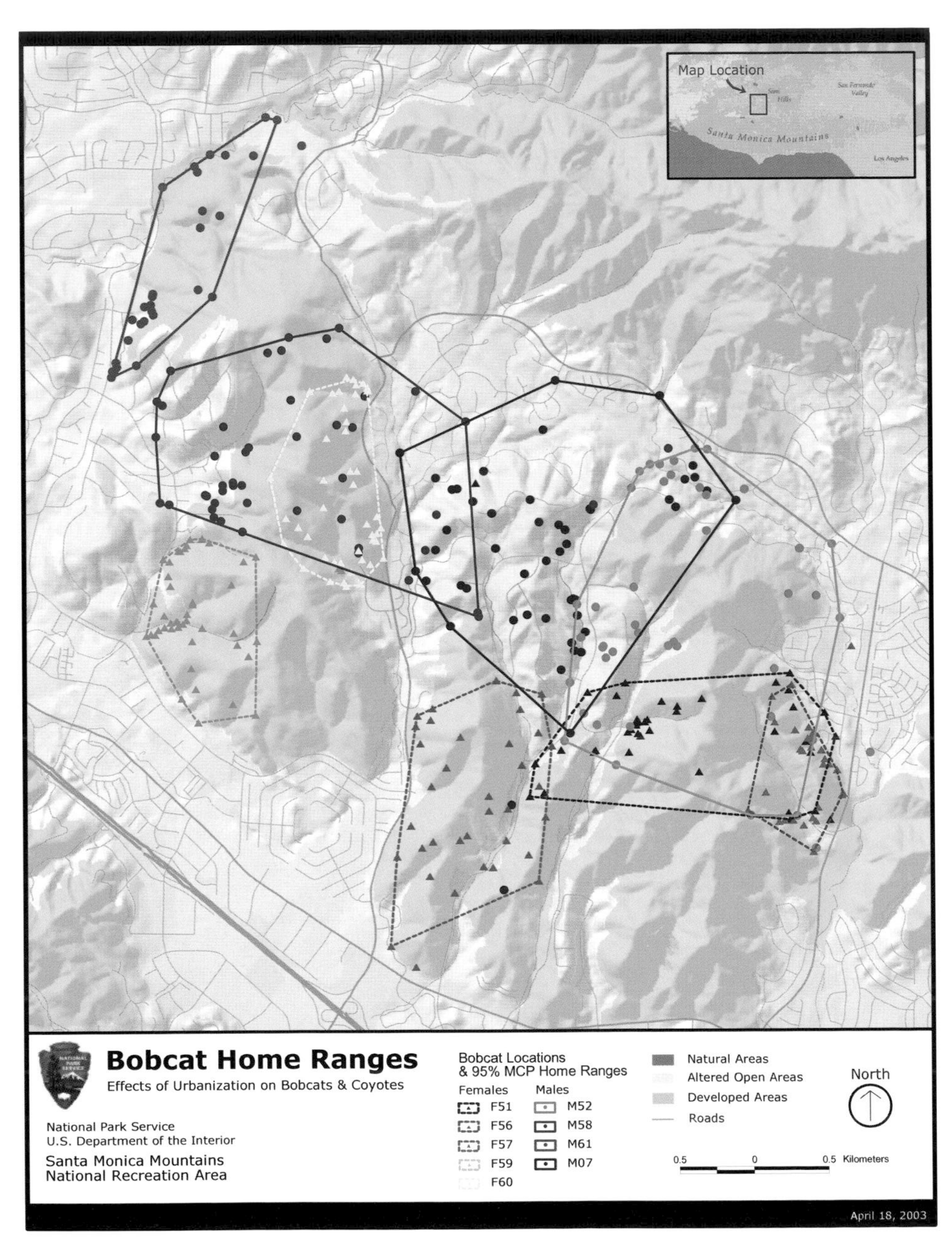

The home ranges of male and female bobcats relative to land use in the Simi Hills area of Ventura County, California. Altered open areas include low-density residential housing, golf courses, and vegetated patches or strips.

Wildlife researchers found the unexpected when they studied how bobcats and coyotes cope with the loss of natural environment in and around the Santa Monica Mountains National Recreation Area in California. Researchers had captured fifty bobcats and eighty-six coyotes, attached radio collars, and monitored their movements. Using GIS technology, the researchers entered the locations into a database displayed on GIS maps. Ecologists have long known that carving up natural areas with roads, housing tracts, and other development can threaten wildlife populations. Yet a study of these two species in the hills west of Los Angeles found survival rates similar to those of bobcat and coyote populations that lived in areas unspoiled by development. Researchers suspect that the relatively high survival rate may have resulted from the mild climate, plentiful prey, and absence of hunting and trapping. At the same time, rodent poisons and vehicle collisions caused a significant number of bobcat and coyote deaths, indicating the pervasive effect of humans and development on wildlife even within the national recreation area. Both species shifted their activity from day to night, particularly in developed areas, perhaps because fewer people are out then. And the animals ranged mostly in natural areas. Researchers concluded that they need to learn more about the requirements of carnivores in developing areas and educate the public on the need to protect nature reserves and use rodent poisons sparingly and correctly.

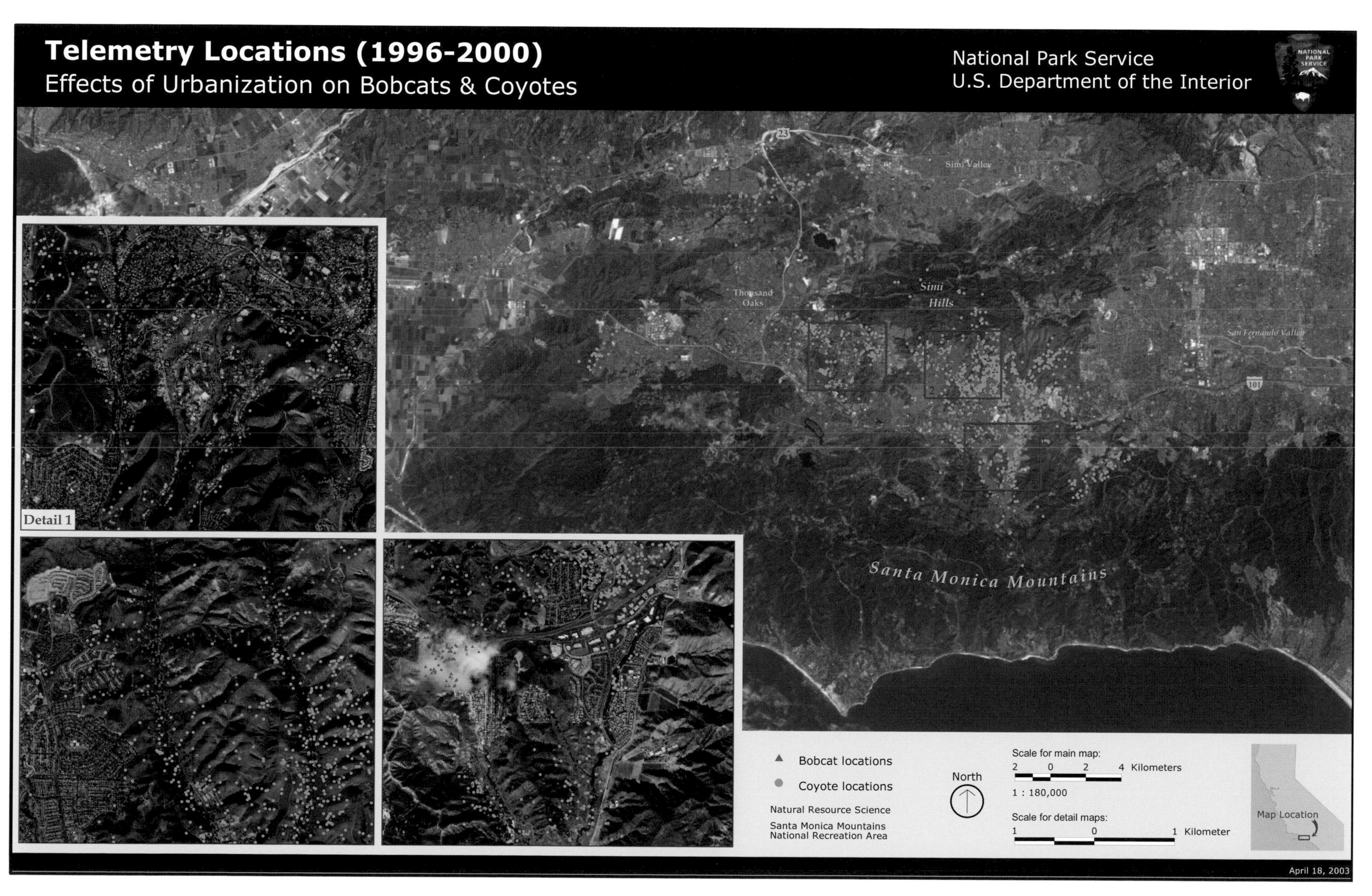

Telemetry locations of radio-collared bobcats and coyotes, 1996 through 2000.

Hawksbill Turtles of Hawaii

Lynette Spujt and Byron Iida excavate a hawksbill nest on Kamehame Beach. Photo: NPS–HAVO Turtle Project.

An army of volunteers using tools of observation, science, and technology is helping the National Park Service protect and nurture a rare turtle in the Pacific Ocean. Researchers in the United States have documented nesting hawksbill turtles only in southern Florida and Hawaii. At one time, these creatures cruised the sea in relatively large numbers. But loss of nesting habitat, predation, and poaching for their shells have reduced turtle populations to critically low levels.

The turtle nests on ten beaches on the island of Hawaii, various beaches on the Maui coastline, and one beach on Molokai. On the island of Hawaii, Hawaii Volcanoes National Park protects the beaches where ten to twenty turtles lay their eggs in a nesting season that lasts from late May to December. A female hawksbill deposits an average of 175 eggs per clutch. She uses her flippers to cover the nest with sand and returns to the sea while the eggs incubate. Two months later, the hatchlings emerge from the nest as a group and scramble to the waves. These tiny creatures face mortal danger from mongoose, nonnative plants, artificial lights that lead them away from the water, traffic, and recreation on the

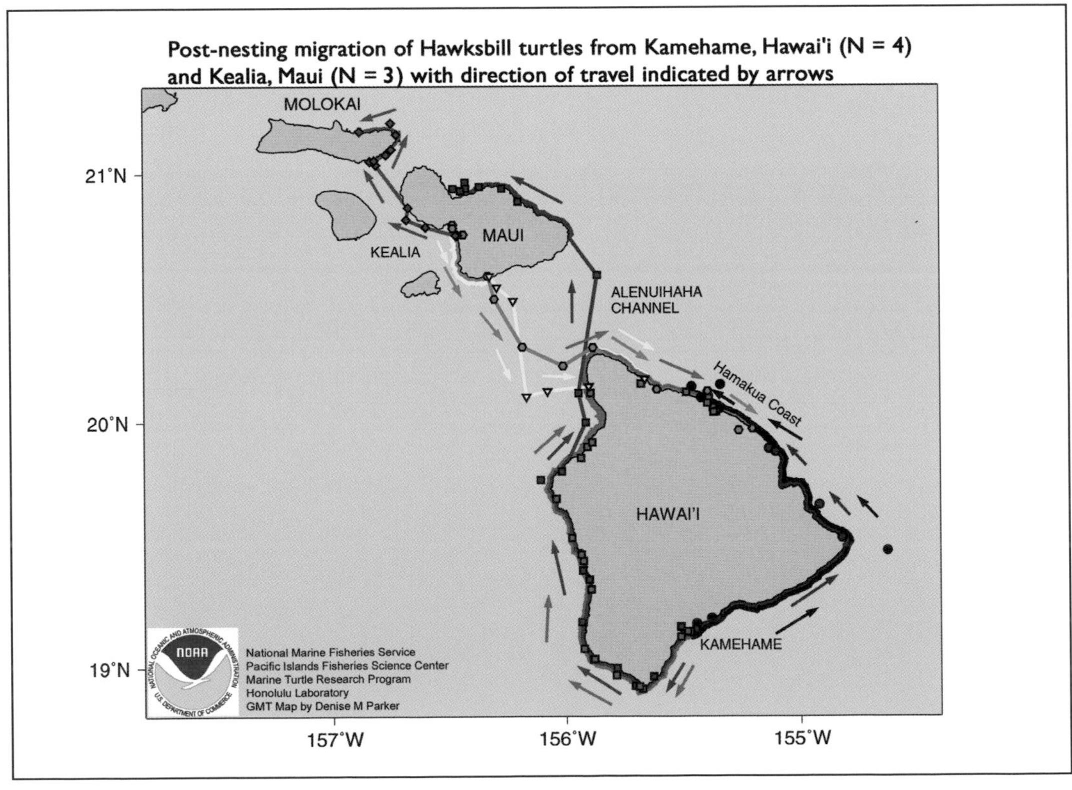

beaches. After depositing between one and six clutches of eggs in a season, the female is not expected to nest again for two to six years.

The research project takes place within the national park and on county, state, and private lands. Volunteers sometimes help hatchlings reach the water. Volunteers also handle and tag turtle flippers for identification, and excavate nests to rescue stranded young and determine nest success. They also trap and kill introduced predators such as mongooses, rats, pigs, and feral cats to protect turtle eggs and hatchlings.

To determine postnesting migration and resident foraging habitat of these turtles, the National Marine Fisheries Service attached satellite transmitters to a few turtles. Data is collected and processed via satellite telemetry. The fisheries service analyzes this information and produces maps. These maps, along with photographs, field data, and scientific research, help the park service and fisheries service develop long-term plans to prevent the hawksbill from disappearing from the earth.

Volunteer Stacy Sauter prepares to mark a nest site. Hawksbills lay an average of 175 eggs per clutch. Photo: NPS–HAVO Turtle Project.

Left: Volunteers identify a postnesting hawksbill as she returns to the ocean. Photo: Larry Katahira, NPS.

Right: A female hawksbill prospects for a nesting site at Apua Point in the national park. Photo: NPS–HAVO Turtle Project.

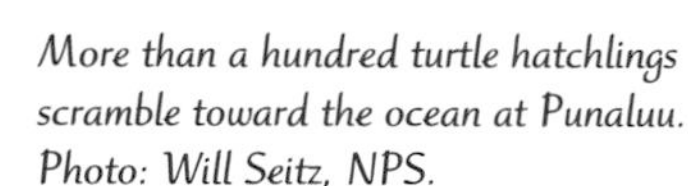

More than a hundred turtle hatchlings scramble toward the ocean at Punaluu. Photo: Will Seitz, NPS.

Volunteer Marcie Matsuo with an approximately 180-pound turtle with a carapace length of about 85 centimeters. Photo: Larry Katahira, NPS.

A hawksbill hatchling heads to the ocean after being rescued from its nest at Keahou Beach in the national park. Photo: Will Seitz, NPS.

Lynx in Yellowstone

Lynx in the wild. Researchers have confirmed that lynx are in Yellowstone.

Science and technology recently confirmed the presence of the elusive Canada lynx in Yellowstone National Park for the first time in decades. Now, the park service is using traditional and modern tools to monitor and protect the shy creature in its ancestral homeland. Wildlife biologists began a parkwide study of the lynx after the United States listed the feline as an endangered species in 2000. In 2001, park biologists collected hair samples that turned out to be from lynx. In 2003, DNA from hair and fecal samples identified a female lynx and her kitten in central Yellowstone. The discovery of a female with her offspring indicated the animals were not just passing through but actually lived in the park. The park service and its partners hope the new evidence bodes well for the future of the lynx in Yellowstone, where sightings of the wild cat date at least to the early 1900s. The Audubon Society found no sign of the animal during a backcountry outing of two hundred members in 1926. Yellowstone last listed the lynx as a "rare native" in the late 1960s, and wildlife biologists feared the species had disappeared from the park. Lynx are difficult to identify by sight alone because they typically travel alone and resemble their cousin the bobcat. The park service now uses GIS technology to identify and study areas of the park where lynx populations might exist and even grow. GIS specialists created a digital model of lynx habitat in the Yellowstone region, using vegetation criteria based on information from biologists who studied the animal in the Rocky Mountains. The map showed the park and its partners where to focus their lynx preservation efforts. The information also helps biologists identify park activities that might harm the animal or its habitat. In one example, GIS analysis showed how ongoing road construction affected lynx habitat at the east entrance of the park.

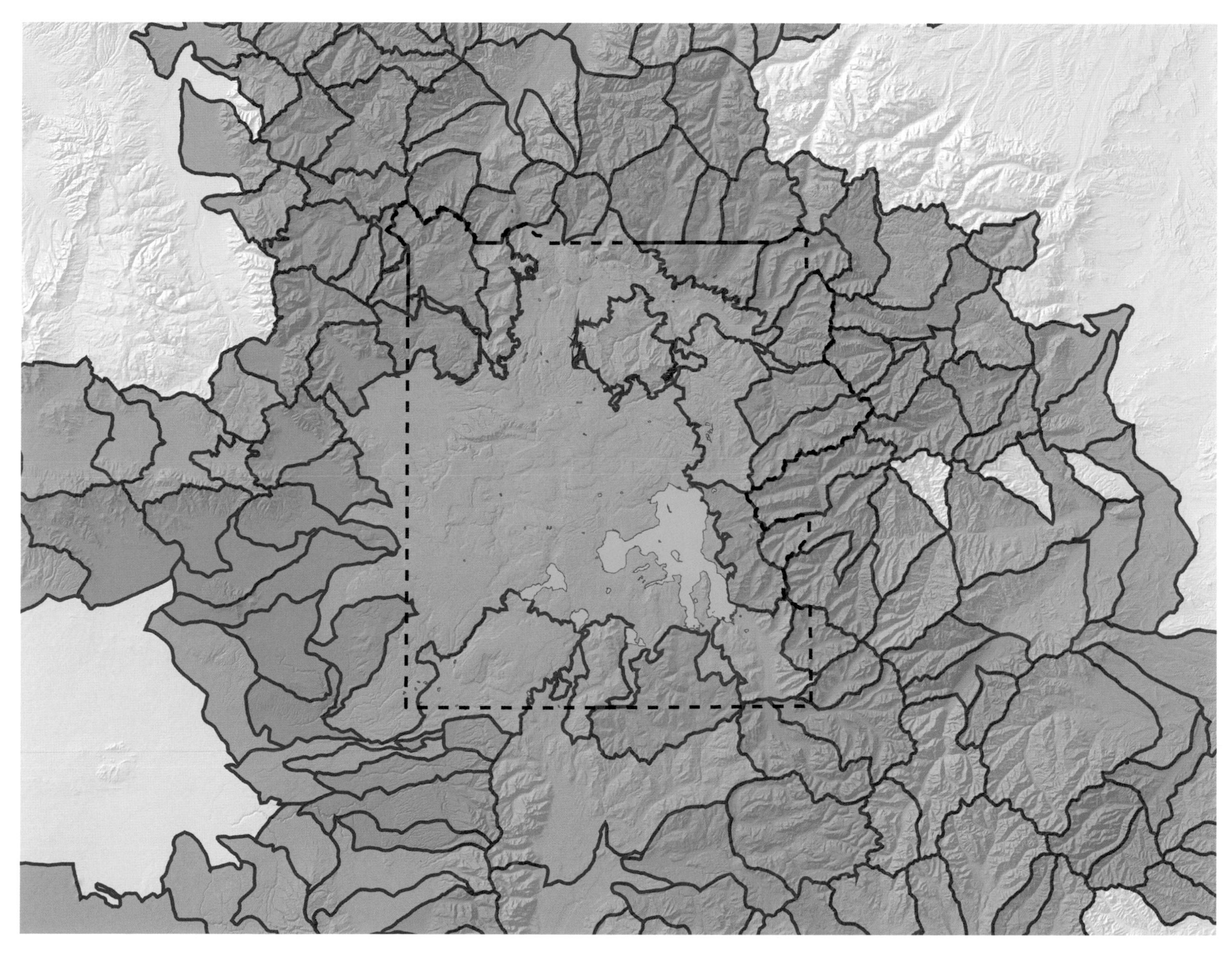

GIS specialists created a digital model using vegetation criteria to display suitable lynx habitat (with blue borders and shading) in and around Yellowstone National Park. The black dashed line indicates the park's border.

A Pilot Geographical Information Systems Assessment of Rocky Mountain Bighorn Sheep Habitat in and around Great Basin National Park, Nevada

Bighorn sheep are getting help from high technology, including GIS, as wildlife biologists try to prevent a symbol of the West from vanishing from Great Basin National Park in Nevada. Today, fewer than a dozen Rocky Mountain bighorn survive in the rugged Snake Range in and around the park, where hundreds of the majestic creatures once roamed. Early on, wildlife biologists tried to bolster the herd by adding bighorn sheep from other areas. Those efforts failed, and the herd continued to dwindle even after the creation of the national park in 1986. In 2001, wildlife biologists used GIS technology to see if the loss of sheep habitat played a role in the herd's woes. Using GIS technology, researchers evaluated whether the herd had enough room to survive and grow.

The resulting GIS map showed the sheep still had plenty of room in general but barely enough to bear and raise their young. Bighorn ewes and their newborn need areas with plenty of grass and water during the spring lambing season. They also need open areas where they can spot mountain lions and other predators and escape pursuit over rocky slopes. One GIS map layer identified the southerly slopes where the snows melt early and provide grasses for forage. Another layer showed the availability of water, and a third identified types of vegetation that indicate open areas and offer protection from predators. From this information, wildlife biologists and fire managers are using GIS to plan remedies, such as prescribed burns and thinning of forests, to restore suitable lambing areas in hopes that the herd will one day thrive again in its historic range.

A Rocky Mountain bighorn ram on the rocky slopes of Wheeler Peak.

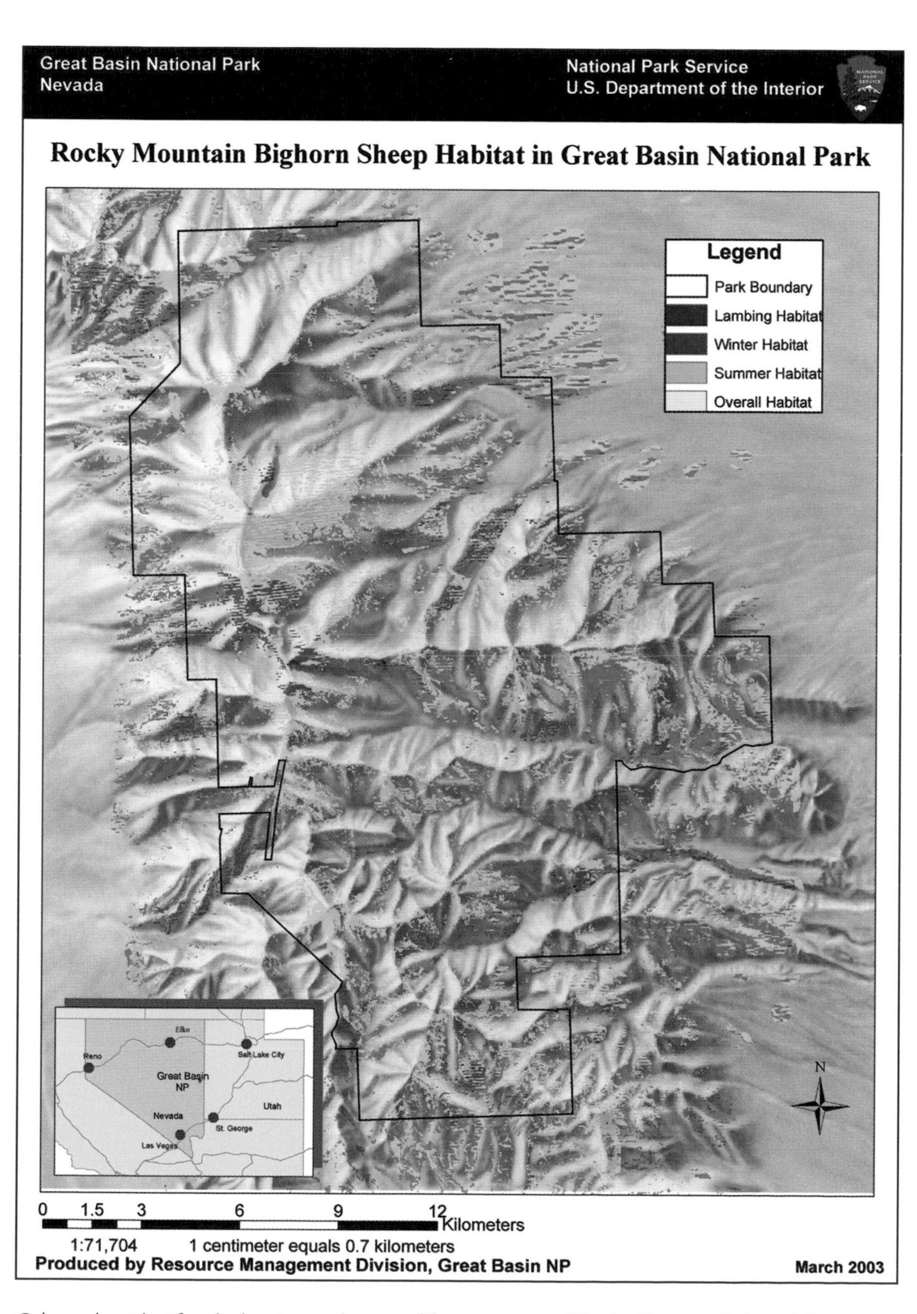

Color coding identifies the location and extent of four categories of Rocky Mountain bighorn habitat (lambing, winter, summer, and overall).

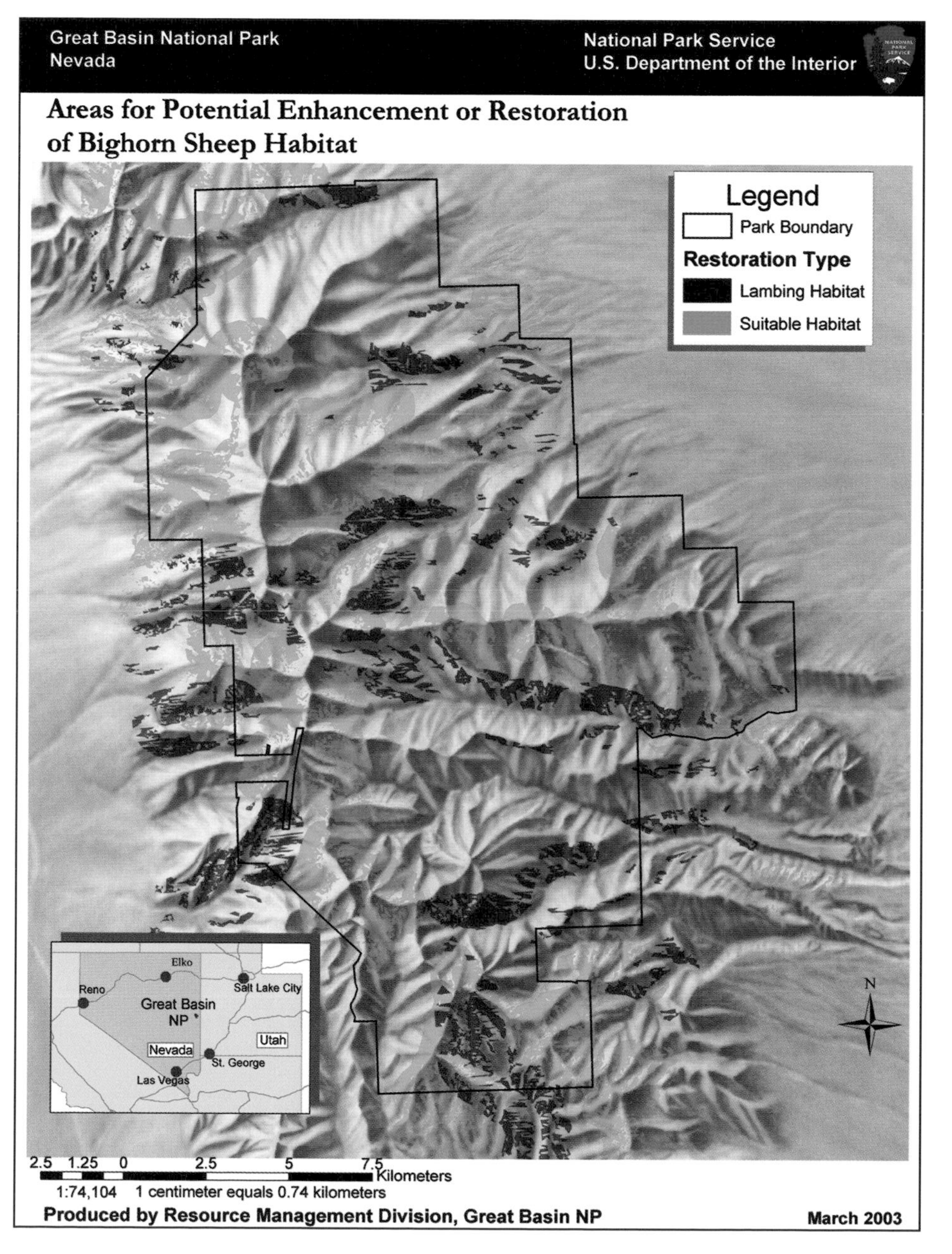

Efforts to improve the herd's prospects are aided by GIS-generated maps like this.

California Condor Viewshed Analysis

Tagged California condors in their release pen not long after their 2003 arrival at the monument.

A GIS map has given researchers at Pinnacles National Monument in California a bird's-eye view of what endangered California condors see from their release pen atop a ridge. The monument needed a way to deliver water and lead-free animal carcasses to the condors, which arrived in the summer of 2003. With a wingspan of ten feet, the largest of any North American land bird, condors in the wild fly long distances in search of carrion. Wildlife biologists wanted to monitor the behavior and health of the six captive-bred immature condors and one older "mentor" condor from a nearby observation post. At the same time, researchers did not want to be in the line of sight of the condors as they returned to the wild. To solve the problem, the park service collected visual data and created a GIS map. Orange areas on the map represented what these naturally curious condors would view from their release pen. The map showed that condors would not see a planned trail to the release pen and observation post. By making themselves scarce, researchers hope the birds will learn to hunt and socialize in this region for the first time in decades and help make the reintroduction program a success.

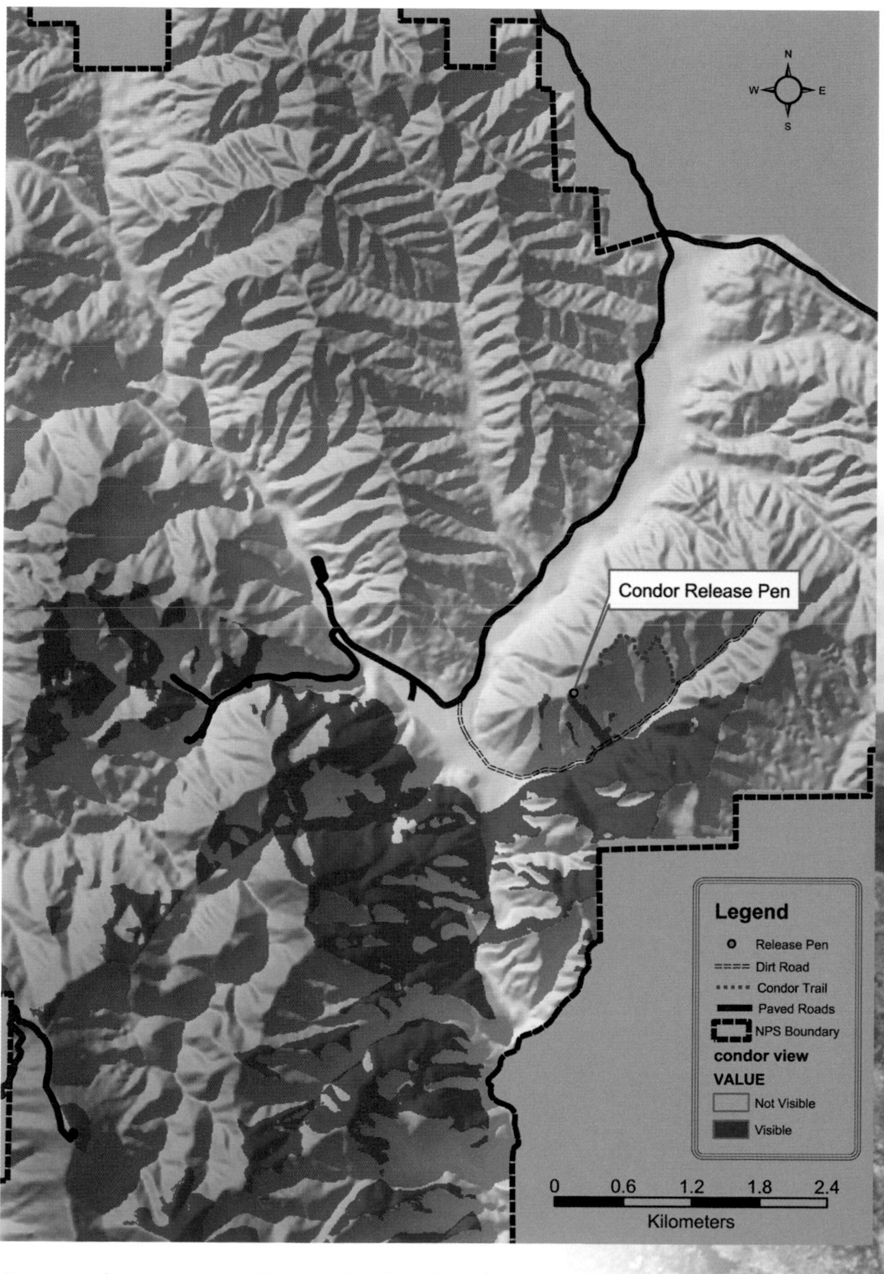

Magenta indicates areas viewable by condors from their release pen in relation to an existing trail and new ATV trail.

The monument's "High Peaks" as seen from the condor release site, looking west.

CHAPTER 11

The Big Picture: Ecosystems

If future generations are to remember us with gratitude rather than contempt, we must leave them more than the miracles of technology. We must leave them a glimpse of the world as it was in the beginning, not just after we got through with it.

President Lyndon B. Johnson, upon signing the Wilderness Act, 1964

Ecosystems are as varied as the meadows, glaciers, rivers, canyons, and other resources that make up the National Park Service landscape. The primary resource of an ecosystem might be a lake, forest, or desert, or a particular endangered plant or animal. This makes ecosystems hard to define. They often include complicated relationships between physical, biological, and cultural resources. And federal law requires the National Park Service to preserve ecosystems that often extend beyond protective park boundaries. Within this realm, GIS is a great tool to understand ecosystems and their importance in our world today.

GIS software has changed the way the National Park Service presents information internally and to the public. Scientists have long known there is no substitute for information and data about resources. But it is another matter to display scientific findings in a simple way that makes common sense to policy makers and park visitors. GIS meets this challenge with colorful maps and posters packed with information that can illustrate the extent of an ecosystem and ease the job of describing it.

The mapping function of GIS illustrates the relationships between individual parts of the ecosystem, such as a particular wilderness and the wildlife found there. This high-tech tool does simple jobs, such as measuring areas and distances, or more complex tasks involving many variables within an ecosystem. GIS software can produce models that show the slope of land, precipitation, or kind of soil, and help us understand their effects on the ecosystem. Geographic information systems can illustrate the relationship of an ecosystem to other areas and let researchers monitor changes that take place over time. Animations or time-series maps offer convincing, understandable evidence when these changes happen.

This wealth of information has helped the National Park Service do a better job as it preserves our parks. With help from GIS, park managers can study and react to potential environmental threats against a park that come from the outside. Using GIS, a park manager might propose extending park boundaries to include resources deemed critical for the health of an ecosystem. More frequently, it helps the National Park Service monitor ecosystems as part of its goal of preserving our natural wonders.

2002 Exotic Plant Management: Chemical Control

President Theodore Roosevelt designated Devils Tower as the nation's first national monument in 1906, recognizing the rock formation in the Black Hills of Wyoming as one of the most remarkable peaks in the country. Yet noxious weeds and other exotic plants in the monument have disrupted the natural process of fire, hydrology, and the renewal of carbon, water, nitrogen, and other nutrients. The weeds also hurt native plant and wildlife populations. The park aims to control these harmful plants to protect the monument, prevent their spread, and preserve the natural view for visitors. The park's weed-fighting arsenal includes the release of flea beetles and other bugs, herbicides, and the removal of exotic plants near development, roads, and trails. It also includes high-tech tools such as GIS software. Using data collected from global positioning systems, GIS maps help the park visualize the distribution of exotic plants and their treatment in the monument.

Workers spray chemicals to control exotic plants at Devils Tower National Monument in Wyoming.

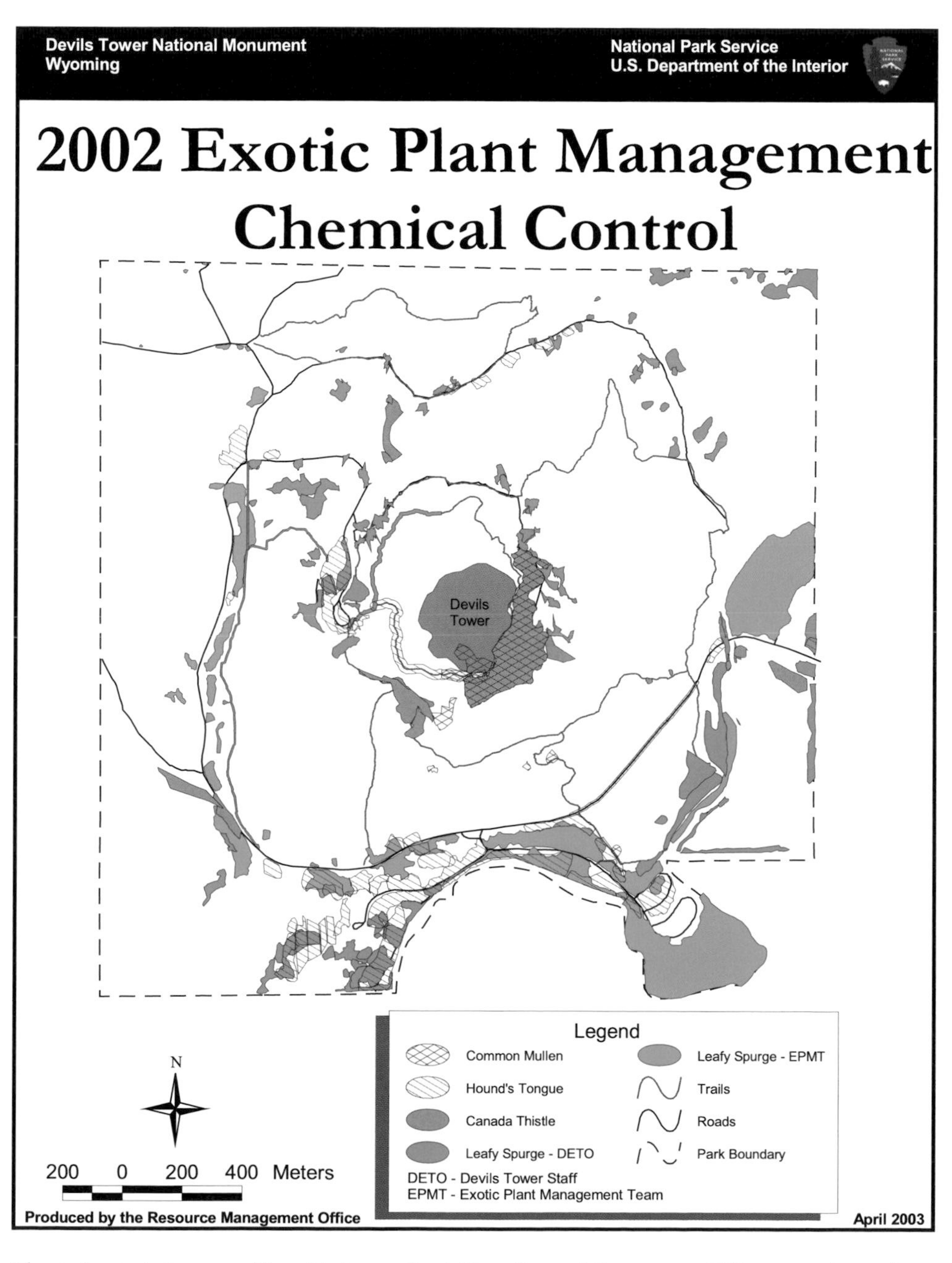

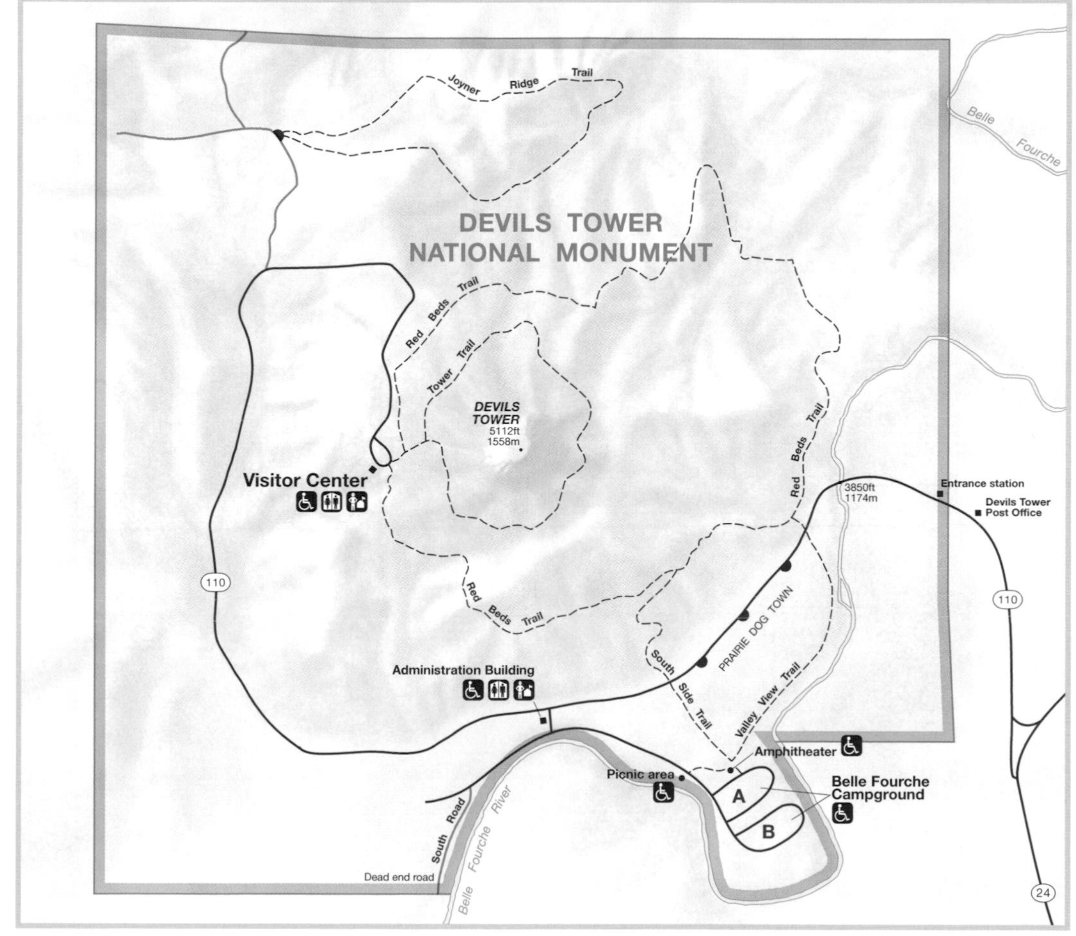

Chemical control of exotic and harmful plants at Devils Tower National Monument in 2002 is summed up in this GIS map.

South Florida Composite Topographic Mapping

The broad movement of water across a nearly flat surface dominates the famous Everglades ecosystem in Florida. The celebrated "River of Grass" flows seaward through grass marshes and cypress forests, and sprawls to a width of thirty miles in some places. Yet for more than a century, governments and individuals have built dams, dikes, levees, canals, railways, and roads that have altered the natural flow, direction, volume, and quality of the river. We have realized that the historic natural flow of the river was the healthy system most beneficial to the plants, animals, and residents. The National Park Service and other parties involved are trying to restore those original conditions, despite the massive changes that have occurred. In this process of restoration, some believe the greatest information challenge is the ability to depict the topographic surface in ways that allow meaningful analysis for local decisions. Maps that show contours of one meter proved largely useless for all but the overview. To meet the challenge, park service geographers created GIS-based maps that provided topographic contours of one-tenth of a foot in the crucial areas. The mapping effort relied on federal, state, and private databases, which varied in extent and accuracy. GIS specialists combined the data sets using automated and hand-drawn contouring to create a regionwide topology for analysis. The affected region includes four national parks: Big Cypress National Preserve, Biscayne National Park, Dry Tortugas National Park, and Everglades National Park. It is crucial to understand the pathways of water from its sources to the sea because dozens of other protected areas are also at the receiving end of this lazy yet tenacious pipeline.

Aerial view of Florida wetlands.

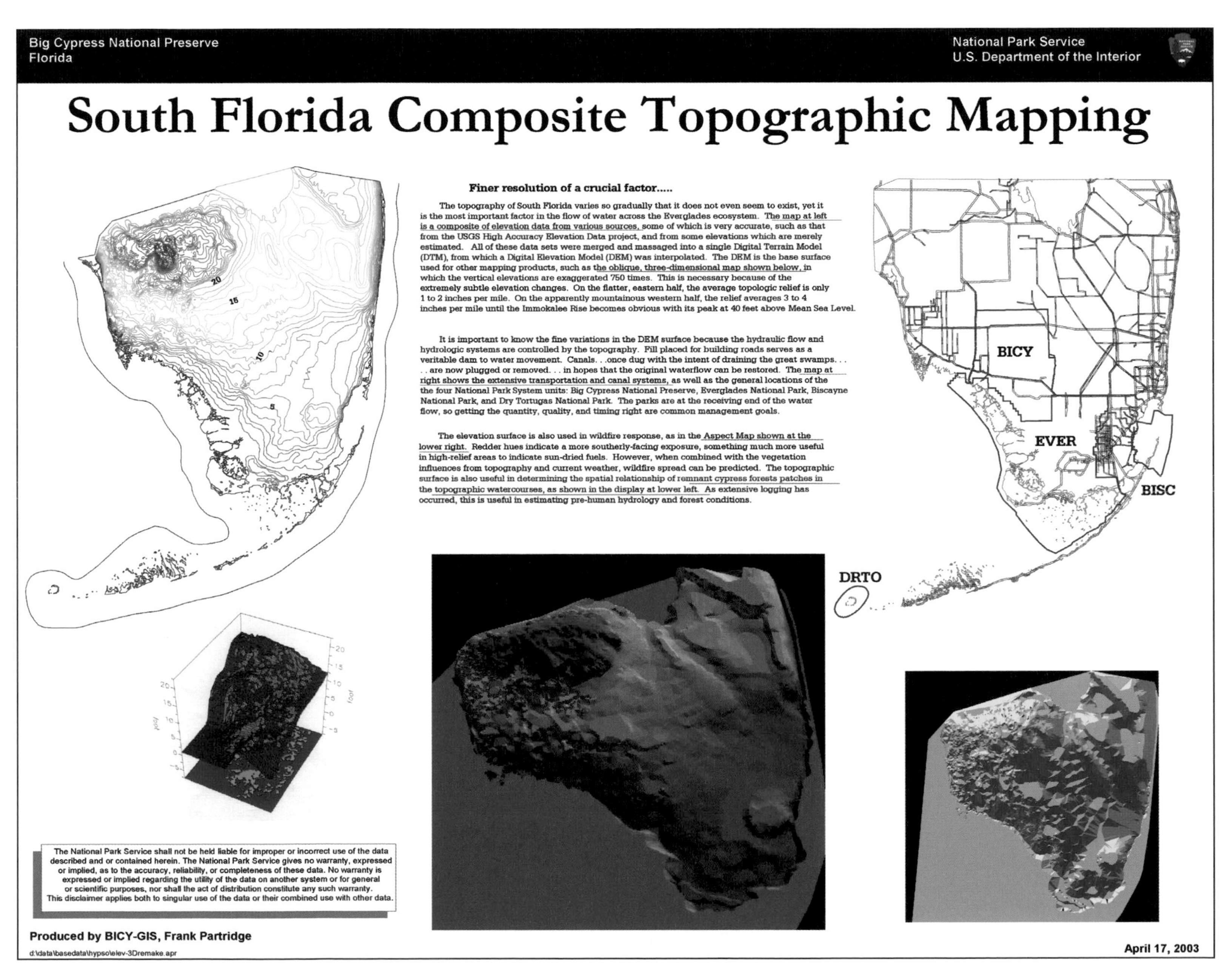

Big Cypress National Preserve
Florida
National Park Service
U.S. Department of the Interior
South Florida Composite Topographic Mapping
Finer resolution of a crucial factor.....
The topography of South Florida varies so gradually that it does not even seem to exist, yet it is the most important factor in the flow of water across the Everglades ecosystem. The map at left is a composite of elevation data from various sources, some of which is very accurate, such as that from the USGS High Accuracy Elevation Data project, and from some elevations which are merely estimated. All of these data sets were merged and massaged into a single Digital Terrain Model (DTM), from which a Digital Elevation Model (DEM) was interpolated. The DEM is the base surface used for other mapping products, such as the oblique, three-dimensional map shown below, in which the vertical elevations are exaggerated 750 times. This is necessary because of the extremely subtle elevation changes. On the flatter, eastern half, the average topologic relief is only 1 to 2 inches per mile. On the apparently mountainous western half, the relief averages 3 to 4 inches per mile until the Immokalee Rise becomes obvious with its peak at 40 feet above Mean Sea Level.
It is important to know the fine variations in the DEM surface because the hydraulic flow and hydrologic systems are controlled by the topography. Fill placed for building roads serves as a veritable dam to water movement. Canals. . .once dug with the intent of draining the great swamps. . . . . are now plugged or removed. . . in hopes that the original waterflow can be restored. The map at right shows the extensive transportation and canal systems, as well as the general locations of the the four National Park System units: Big Cypress National Preserve, Everglades National Park, Biscayne National Park, and Dry Tortugas National Park. The parks are at the receiving end of the water flow, so getting the quantity, quality, and timing right are common management goals.
The elevation surface is also used in wildfire response, as in the Aspect Map shown at the lower right. Redder hues indicate a more southerly-facing exposure, something much more useful in high-relief areas to indicate sun-dried fuels. However, when combined with the vegetation influences from topography and current weather, wildfire spread can be predicted. The topographic surface is also useful in determining the spatial relationship of remnant cypress forests patches in the topographic watercourses, as shown in the display at lower left. As extensive logging has occurred, this is useful in estimating pre-human hydrology and forest conditions.
20
15
10
5
BICY
EVER
BISC
DRTO
The National Park Service shall not be held liable for improper or incorrect use of the data described and or contained herein. The National Park Service gives no warranty, expressed or implied, as to the accuracy, reliability, or completeness of these data. No warranty is expressed or implied regarding the utility of the data on another system or for general or scientific purposes, nor shall the act of distribution constitute any such warranty. This disclaimer applies both to singular use of the data or their combined use with other data.
Produced by BICY-GIS, Frank Partridge
d:\data\basedata\hypso\elev-3Dremake.apr
April 17, 2003

Detailed topographic mapping of Big Cypress National Preserve provides meaningful analysis for local decision making.

Unified Ecoregions of Alaska

Scientists have produced a map that displays the major ecosystems of Alaska for research and planning, and to help protect, understand, and enjoy the values of this vast landscape. The international, interdisciplinary, and interagency effort brought together and expanded on earlier mapping efforts dating to the early 1950s. Ecosystems are identified primarily by their climate and shape of the land, or topography, with minor adjustments based on vegetation and geology. The new map portrays ecosystems as they span international boundaries, including parts of nearby Canada and Russia. Scientists wanted a new map to ease confusion over the need to use and compare several different mapping and classification systems in their research, and to improve communication among agencies. The new map draws on relatively new tools such as satellite imagery, computers, and GIS data and software. Scientists used GIS to analyze these data sets along with digital elevation data, permafrost data, and hydrography, or water characteristics. The map describes each of thirty-two ecoregions according to its major ecological and geomorphic processes, vegetation patterns, dominant wildlife, geologic features, and major climatic patterns. Ecological processes refer to the relationship between organisms and their environment. Geomorphic processes deal with changes in the earth's surface. From the map, researchers and the public can learn about climate, precipitation, the rugged landscape, plants, wildlife, glaciers, and other features. The final product represents the collective wisdom of nearly fifty scientists with hundreds of years of experience in Alaska. The map, with its data set and extensive user guide, gives managers a valuable tool to understand resources under their care.

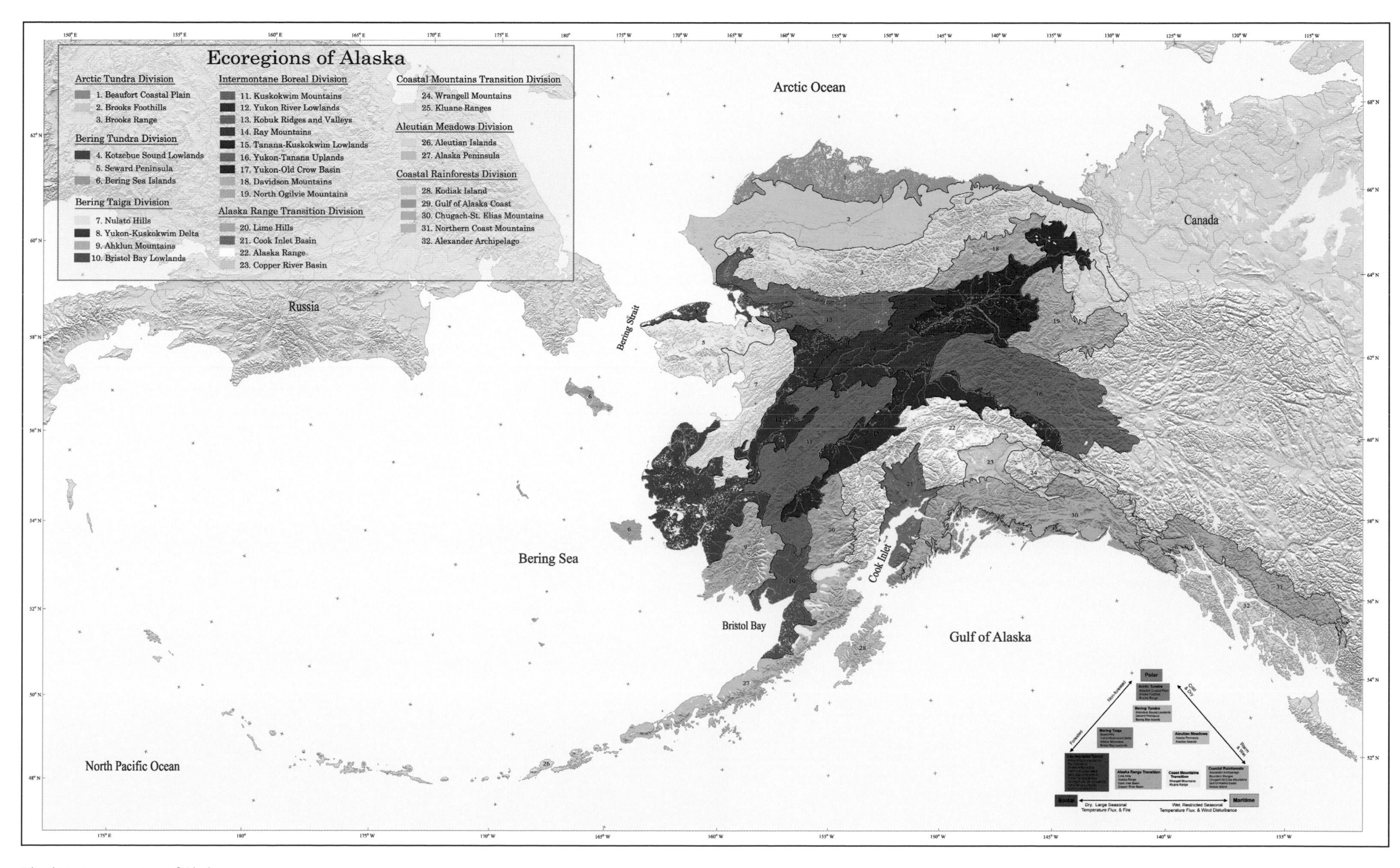

The thirty-two ecoregions of Alaska.

Restoration of Historically Restricted Estuaries, Cape Cod National Seashore, Massachusetts

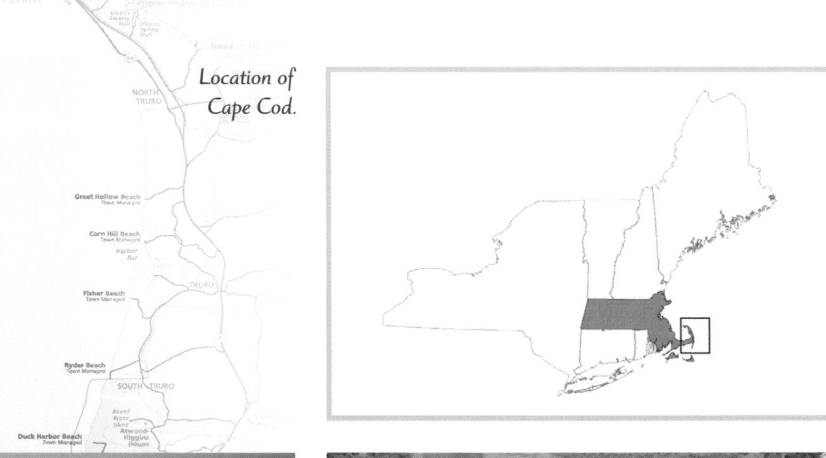

Location of Cape Cod.

Aerial view of seashore.

Aerial photograph of the Herring River in Wellfleet, Massachusetts.

Historical misunderstandings and a lack of appreciation for Atlantic Coast estuaries and salt marshes have led to their widespread draining and development. Federal law created Cape Cod National Seashore in 1961 to protect some of these areas for recreation, conservation, and scientific study. Even so, the National Park Service must overcome an array of obstacles to restore salt marshes damaged by fragmented ownership, development, and lack of knowledge about human impacts. At the national seashore, GIS, global positioning systems, and other technology tools help to restore tidal salt marshes and estuaries at Hatches Harbor, East Harbor, and Herring River. GIS maps display the geographic relationship between vegetation, wildlife, coastal waters, and tidal floodplain elevation, and help specialists predict the effect of tides flowing through culvert openings.

At Hatches Harbor, the National Park Service in cooperation with local, state, and federal agencies is restoring a native salt marsh to a level that will not compromise safety at the nearby municipal airport. GIS specialists are mapping tide heights, deposits of sand, rocks, and other particles, mosquitoes, salinity, flooding duration, and other themes that will help guide restoration strategy over decades. On the Herring River, scientists are restoring natural salt marshes that existed for thousands of years before European settlement interrupted the natural cycle with the placement of dikes starting in the 1700s. By 1910, dikes designed to reduce mosquitoes at the river mouth instead caused most of the original marshlands to disappear. The long history of dikes, drains, and depletion of oxygen in the water helps cause fish kills, reduces fish and shellfish populations, and hurts water quality. The mosquito thrives with fewer predators around. The program to restore the marshland should improve wildlife and fish habitat and leave fewer mosquitoes. GIS technology is expected to help make this happen by modeling solutions to prevent saltwater intrusion into wells and flooding of a nearby golf course. Displaying data and models on GIS maps will play a big part in helping scientists solve these and other restoration issues at the national seashore.

Marshland at Cape Cod.

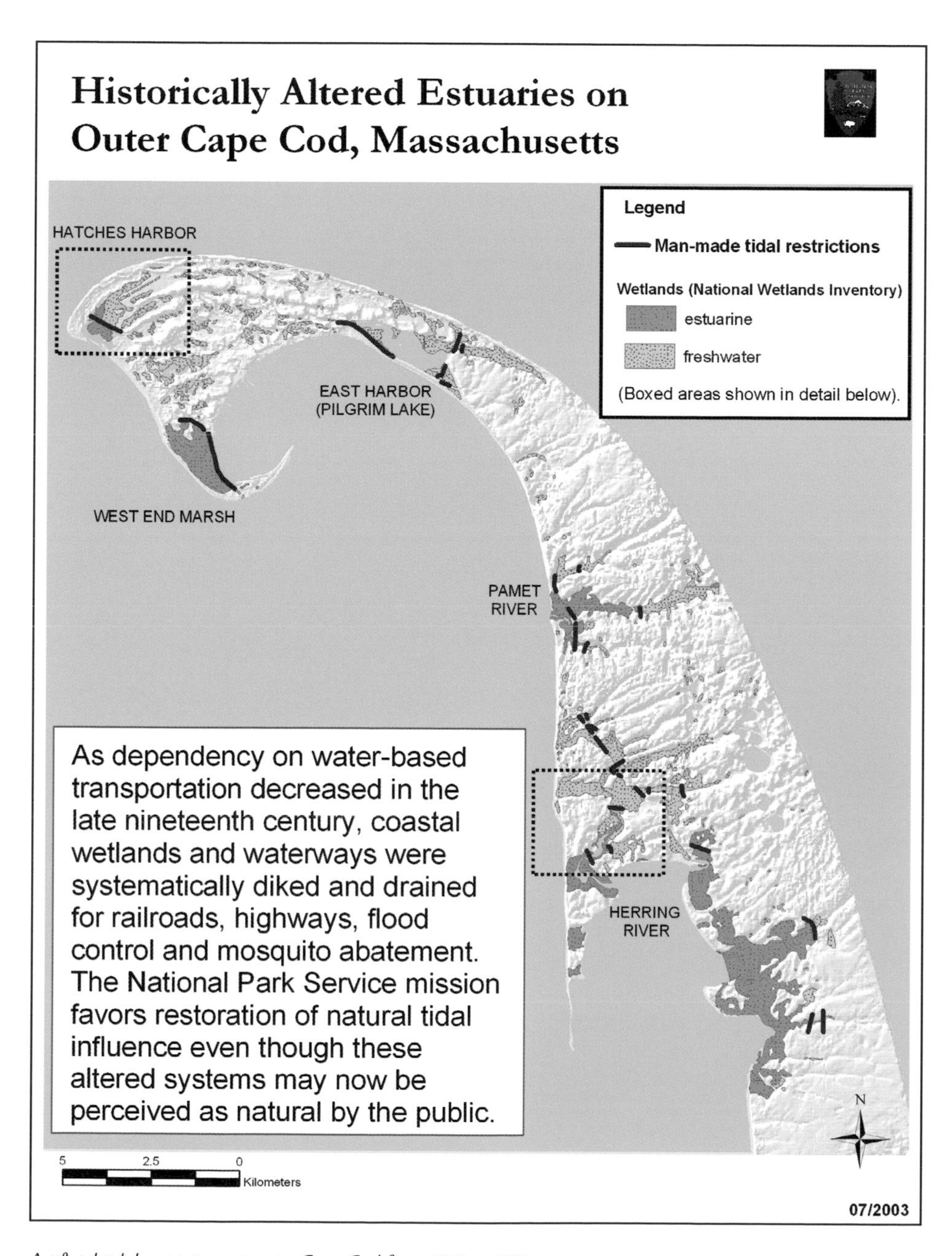

Artificial tidal restrictions at outer Cape Cod from 1871 to 1981.

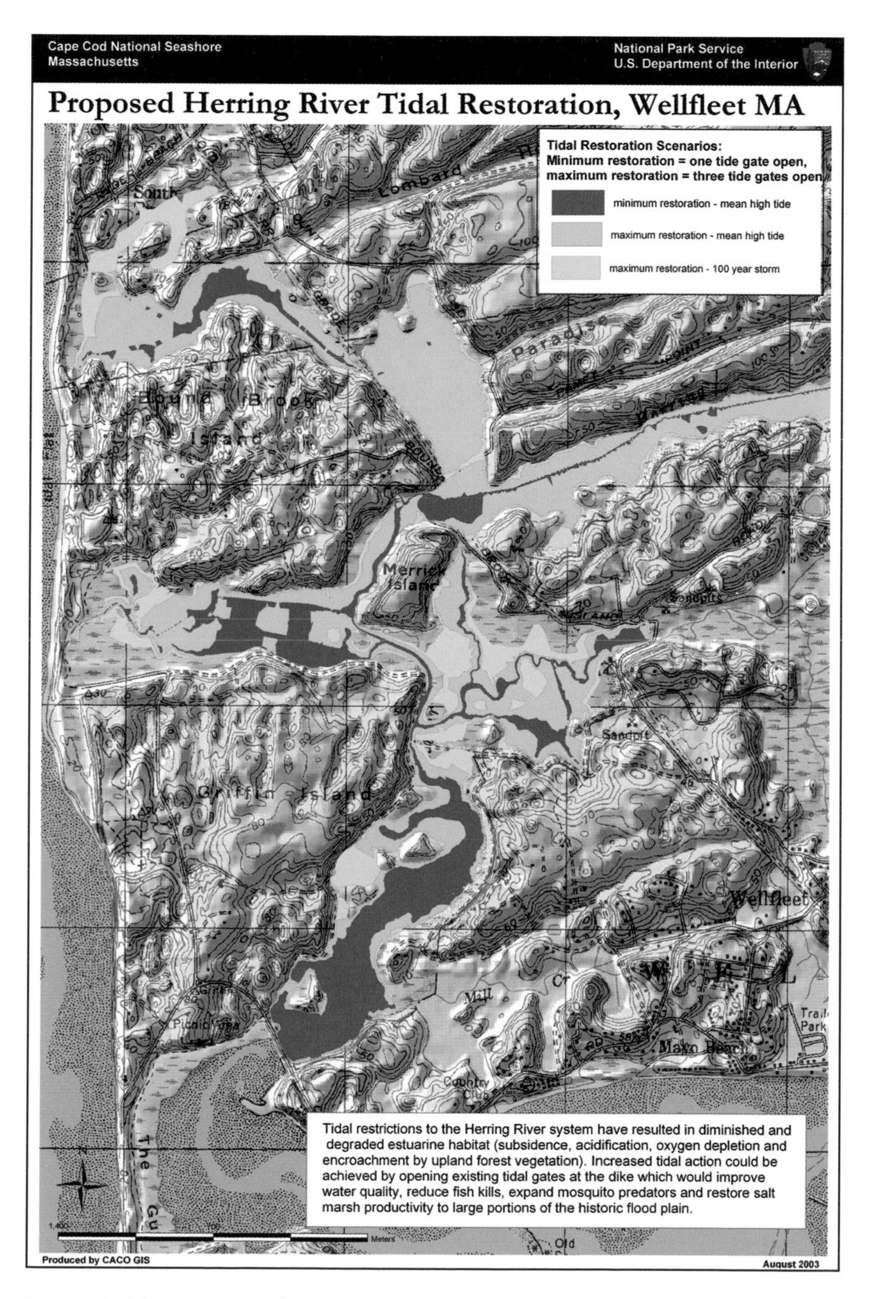

Proposed tidal restoration on the Herring River in Wellfleet, Massachusetts.

Alaska Coastal Resources Inventory and Mapping Program

A high-tech inventory and mapping program in Alaska offers national park managers and the public easy access to resource information about coastal areas for purposes of research, education, management, and preservation. The program at Glacier Bay National Park and Preserve, Sitka National Historical Park, and Klondike Gold Rush National Historical Park displays the information on colorful GIS maps for easy review. The maps display baseline data to help managers evaluate changes in the coastline, determine areas needing special protection, plan for environmental disasters, study historical and archaeological sites, and explain park features to the public. The GIS maps show more than 960 miles of coastline within the three parks and draw from more than 21,000 ground photographs and 320 aerial photographs. GIS technology combines a database, photos, and maps into one easy-to-use visual display to help managers and researchers guide and preserve the parks.

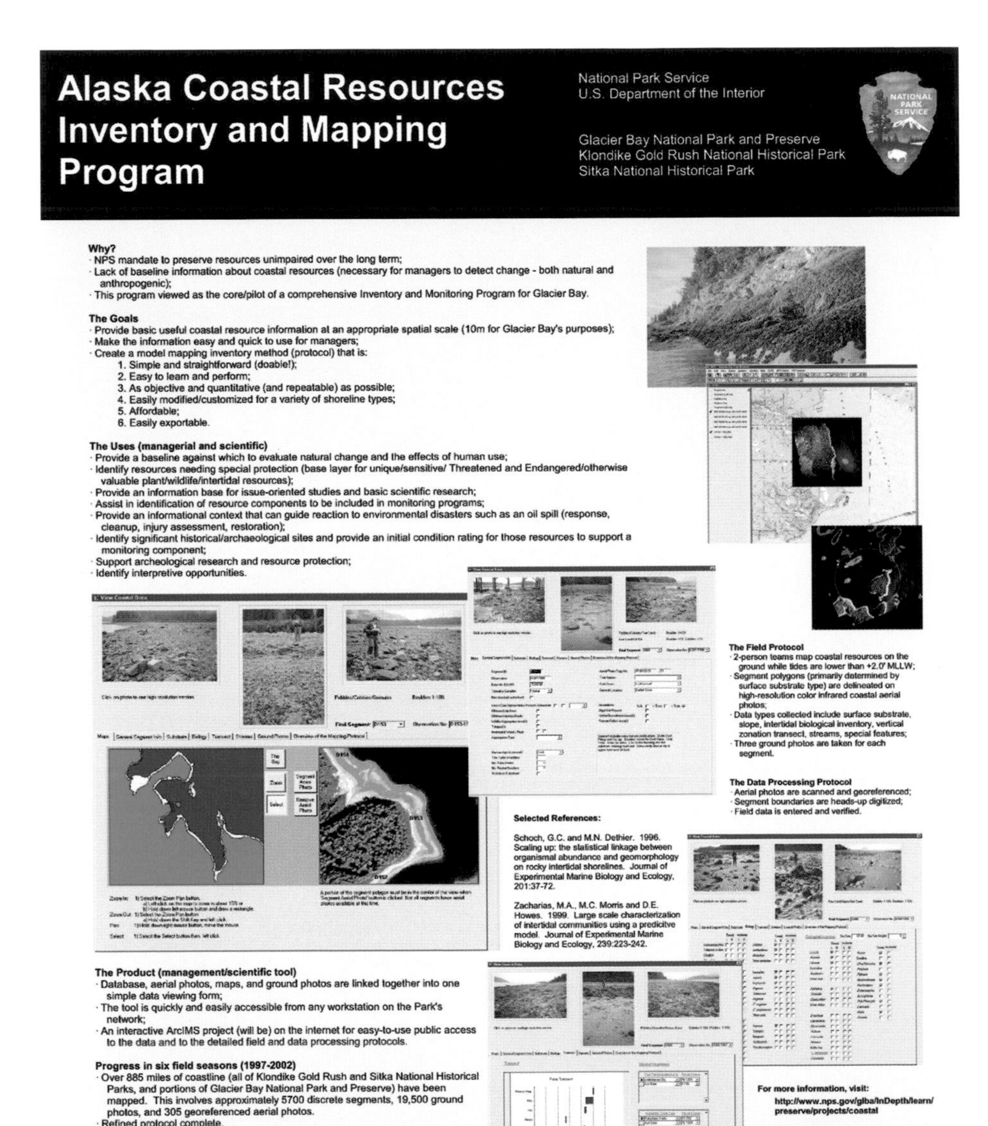

GIS makes coastal data easily accessible by scientists and park managers. This poster combines a database, aerial and ground photos, and maps in a single display. It is also designed for easy public access on the Web.

A representative view of the coastline of Glacier Bay National Park and Preserve at low tide.

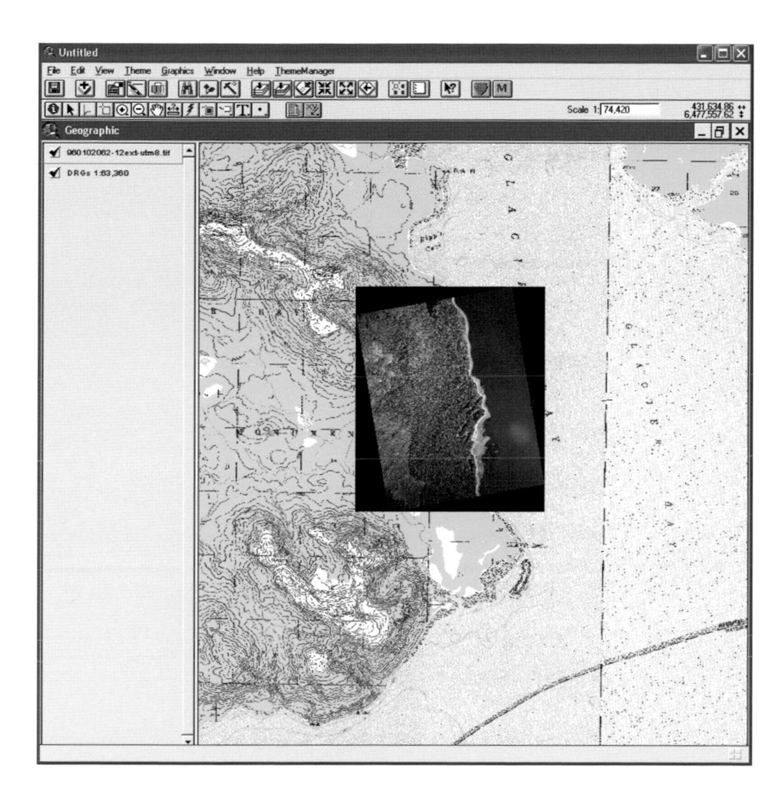

An illustration of an infrared aerial photo georeferenced to a map. Segment polygons are then digitized on the aerial photo to be linked to the database.

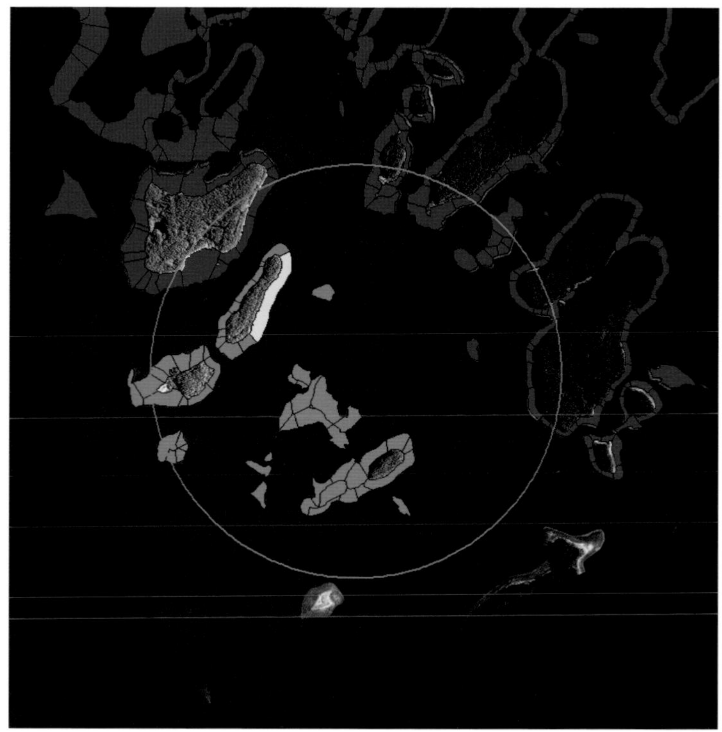

Segment polygons that have been digitized using a georeferenced infrared aerial photograph. Polygons are then linked to their respective coastal resource data in the database.

CHAPTER 12

Setting GIS on Fire: Managing Fire

By forces seemingly antagonistic and destructive Nature accomplishes her beneficent designs, now a flood of fire, again in the fullness of time an outburst of organic life.

John Muir

Wildland fires burn across hundreds of thousands of acres of national parklands each year. These fires dramatically alter the vegetation and landscape and are a powerful force of change in national parks. The National Park Service aggressively fights and suppresses fires that occur near populated areas, visitor facilities, and valuable natural and cultural resources. However, in many cases, wildland fires are allowed to burn to fulfill their natural role upon the landscapes of the national parks. Wildland fires benefit nature by rejuvenating soils, opening up wildlife habitat, and clearing excess vegetation below the canopy of the forest.

Fire management in national parks today reflects a commitment to public safety and a realization that fire plays an important role in the natural cycle of renewal. When fires burn in our national parks, fire managers turn to an array of tools—from the venerable Pulaski ax dating to the early 1900s to modern-day computers, laptops, and satellite imagery. The park service increasingly relies on a high-tech fire management tool called GIS, which has spread through all levels and geographic areas of our vast national park system. Fire managers use geographic information systems to protect invaluable natural, historical, and cultural resources in national parklands. The technology may play an even larger role in strategic planning and budgeting for fire management.

Mapping has long played a key fire management role in the park service. Maps from nearly a century ago showed the extent of fires within Yellowstone and other national parks. Park service fire managers began to envision the potential of GIS mapping as the technology developed in the early 1980s. By the mid-1990s, GIS proved it belonged in the national arsenal of fire management tools with the development of software programs that help predict fire behavior.

Using the latest GIS technology, fire managers today can combine an almost unlimited number of layers of information onto a single map that displays a fire's relationship to a national park and surrounding area. These individual layers, or themes, can represent vegetation and fuels, historic fire origins and burn patterns, and the location of cabins, campsites, and important habitat and archaeological sites within park boundaries. Firefighters equipped with hand-held computers and laptops can call up these maps and quickly respond while on the fire lines. Using GIS and related technologies such as global positioning systems, digital cameras, and hand-held computers, they can send data on fire perimeters and burn rates to the news media and public via the Internet and to fire managers at distant command centers for near-real-time analysis.

From their analysis, fire managers can predict a fire's intensity, rate of spread, and potential maximum size. Displaying this data on a map helps fire managers determine the placement of firefighters and need for evacuations. Maps of fire histories also show where fire has not occurred at normal or natural intervals. This provides invaluable information about the buildup of vegetation and fuel, and thus the potential for a catastrophic fire and the need for prescribed burns, fuel breaks, and long-term prevention strategies. Today, the challenge is not so much to discover new applications of GIS to fire management, but rather for fire management to keep up with the continuously evolving field of geographic information systems.

GIS: Tool for Fire Managers

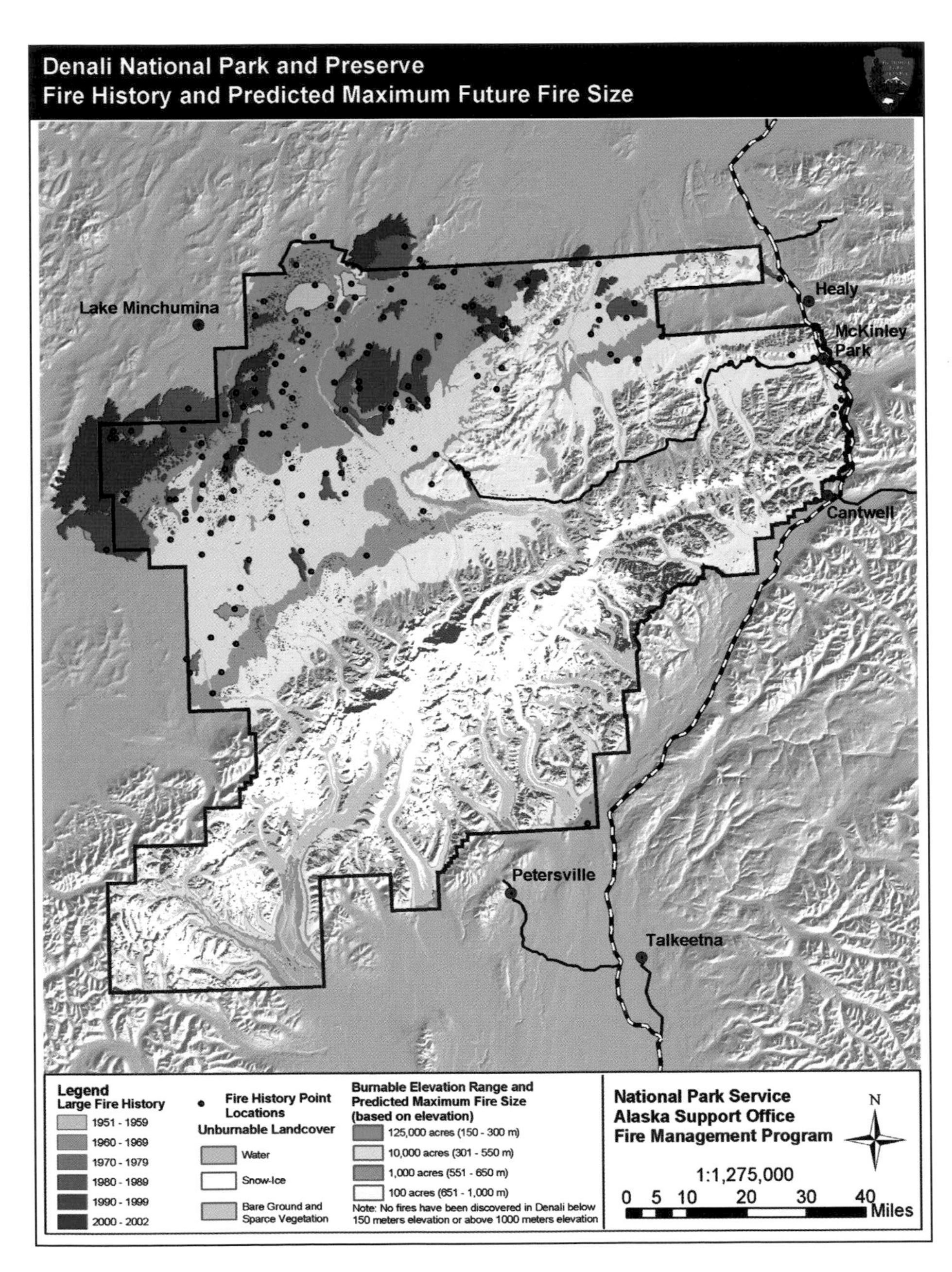

GIS shows the history of fire in Denali National Park and Preserve between 1951 and 2002 and the predicted maximum size of future fires.

The National Park Service integrates GIS with wildfire management across all levels and geographic areas of its vast system. The effective management of a complex process, like a wildfire, requires the interpretation of accurate spatial data. GIS data layers and GIS analysis give fire managers a comprehensive depiction of a park's overall relationship with wildfire. Data layers for vegetation and fuels describe the variable distribution of fuels throughout a park and estimate expected or potential fire behavior across the landscape. Fire history themes, describing the ignition point of all known fires and the total area burned by all known large fires, illustrate the geographic extent of a fire's historic impact on a park. By themselves, GIS themes describing historic fire activity provide a wealth of information to park managers and the public. GIS applications and global positioning system technology also accurately and quickly display fire-perimeter and fire-progression maps to benefit fire managers, fire crews, the media, and the public via the Internet. Park service fire managers welcome and embrace the evolving applications of GIS and related technologies at a time of increased development and other infrastructure in and around parks, as more visitors enjoy these special places.

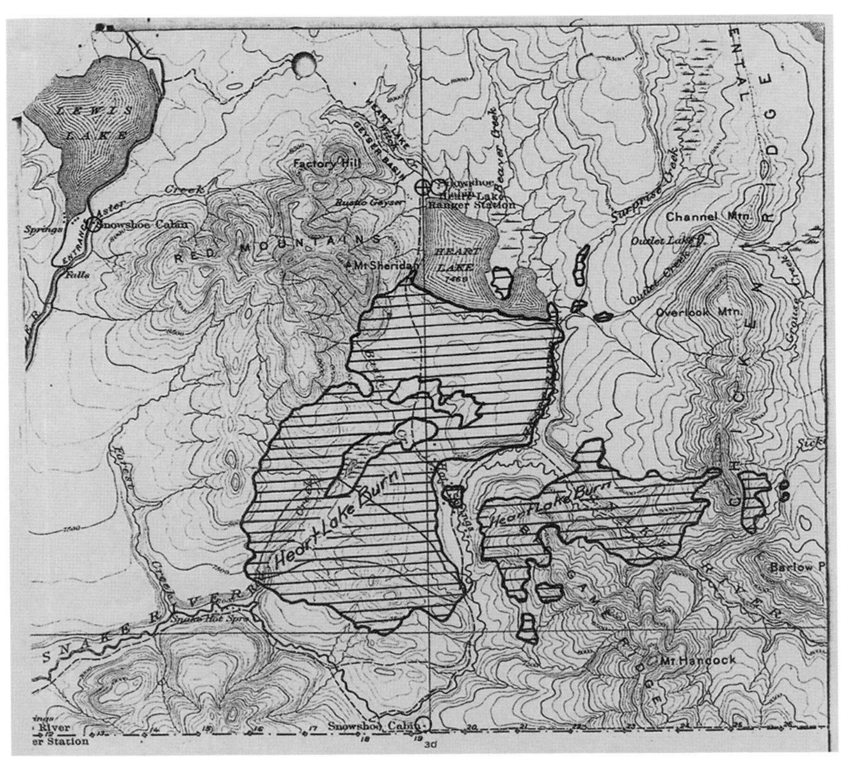

The National Park Service has long relied on mapping as a fire-fighting tool. This 1931 hand-drawn map illustrates the Heart Lake Fire in Yellowstone National Park.

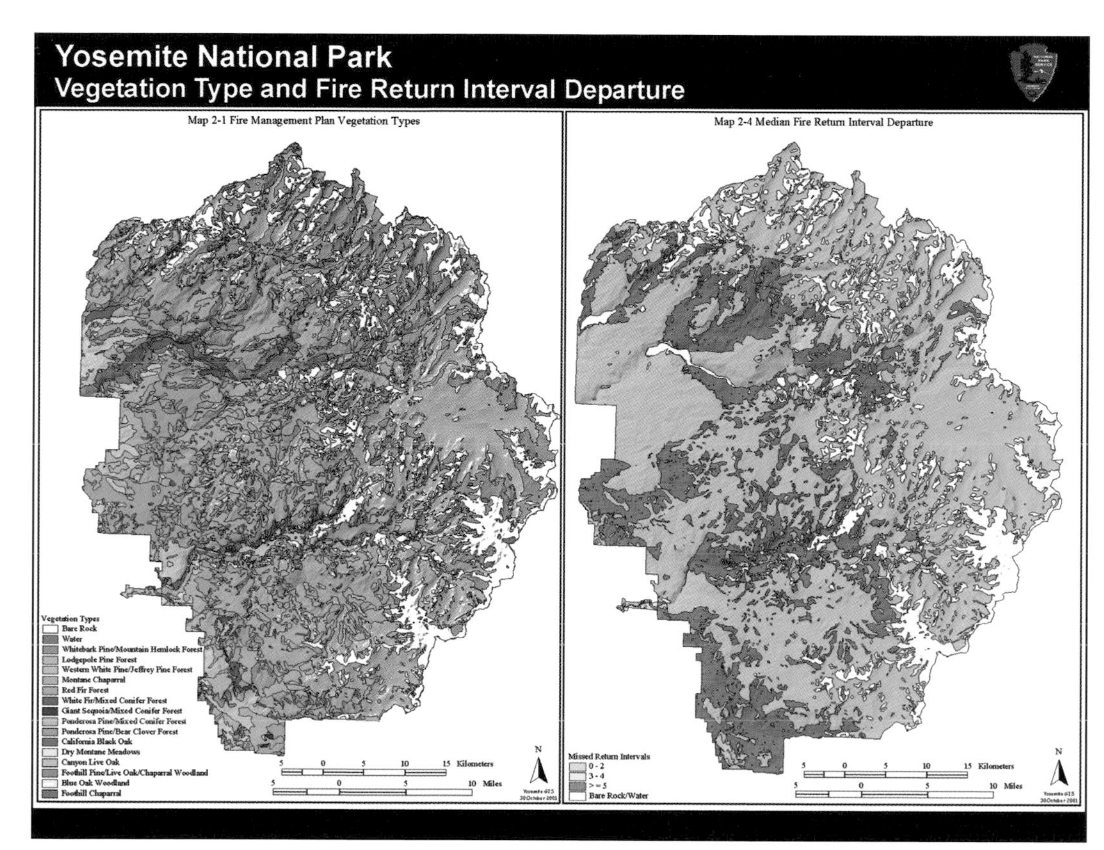

The map at the left identifies the vegetation types of Yosemite National Park. At the right, historic patterns of fire activity and exclusion are displayed in a Fire Return Interval Departure map, a kind of map that shows how much an area departs from normal or natural fire and fuel conditions. Map produced by Kent van Wagtendonk, National Park Service.

Flames move through trees and undergrowth during the Stonetop Fire in Yellowstone National Park in 1999.

The Moose Lake Fire burns in Denali National Park and Preserve in Alaska in 2002.

Using GIS to Support Collaborative, Landscape-Level Fuels

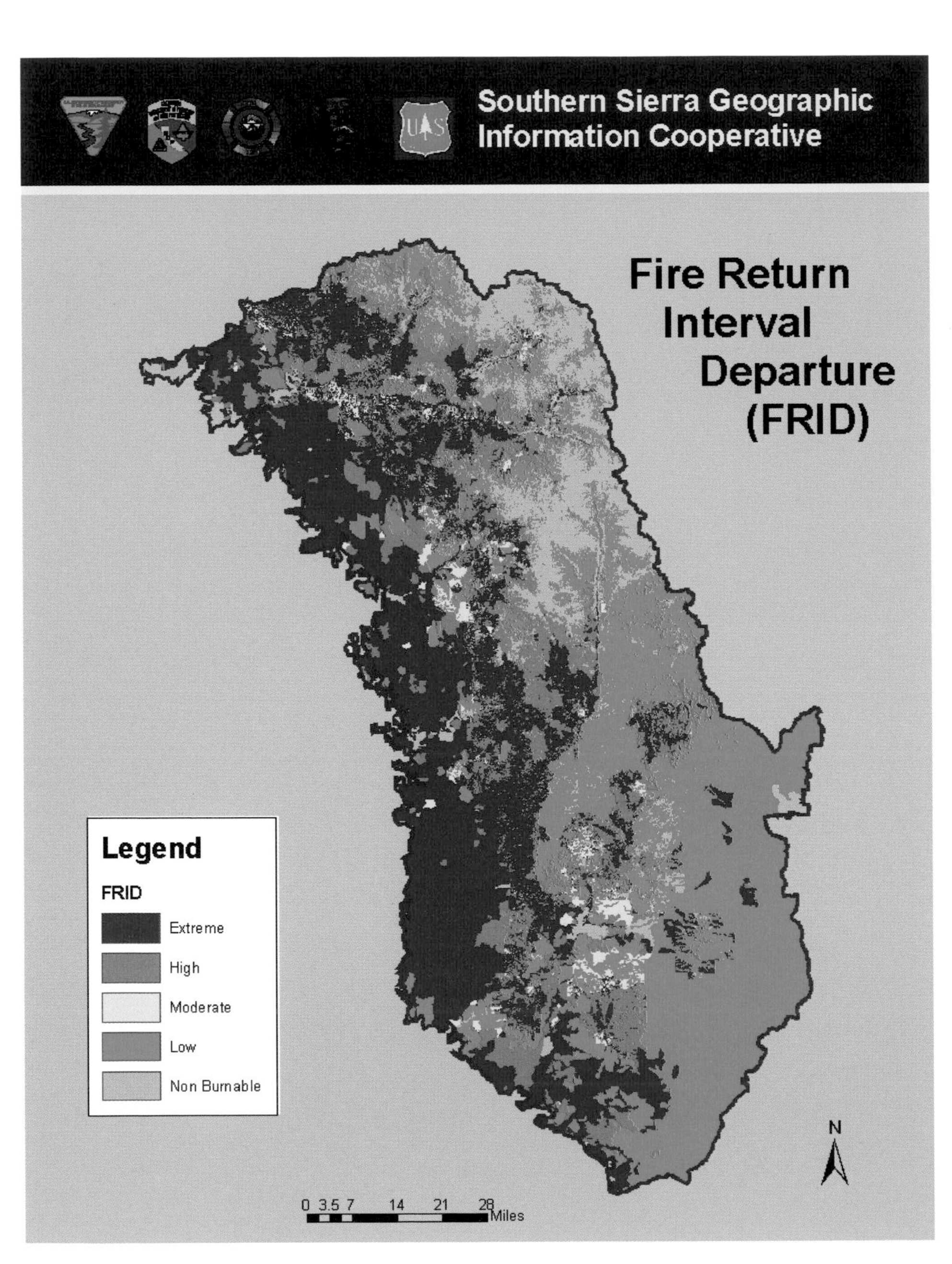

A nationwide increase in catastrophic wildfires led to a program in the Sierra Nevada of California that showed how GIS and Web technology promote fire safety and a healthy ecosystem.

Fire managers started the GIS initiative in 2000 in response to a series of wildfires that burned through rural areas where more people are coming to live. The fires illustrated the conflict between the need to protect lives and property and the need to protect nature, which requires fire to prevent the buildup of dry or dead trees and brush. Historically, the region evolved with naturally occurring, periodic fires that determined its character and maintained its health. But a century of fire suppression led to a dramatic increase of hazardous fuel. To deal with the threat, the park service and other federal, state, and local agencies collaborated on a fire management plan for 4.7 million acres in the southern Sierra Nevada, including Sequoia and Kings Canyon national parks.

These agencies previously collected, coded, and measured fire management data in slightly different ways, making collaboration difficult. But by using GIS and Web technologies, the agencies combined complex data and completed common analysis. This information is now available on a map-based Web site. GIS maps included a range of data themes, including fire behavior, historic fire origins, intervals between fires, projected flame lengths, fire threat to rural communities, and firefighter safety. Moreover, the maps helped fire managers set priorities for areas that need controlled burns or brush removal. The program has become a national prototype to show how GIS and Web technologies help agencies collaborate on fire management planning.

This Fire Return Interval Departure (FRID) map of the approximately 4.7-million-acre southern Sierra Nevada in California illustrates the departure from the average interval of time between naturally occurring fires across the landscape. The FRID system assigns a number to represent the number of fire return intervals that have been missed. For example, if an area with an average fire return interval of 25 years has not experienced fire in 100 years, then the area has missed four fire return intervals. The absence of fire in an area can lead to an unnatural buildup of fuels and result in more intense fires once they finally occur. Fire managers use FRID to assess landscapes and identify where prescribed burns or other fuel-removal remedies might be necessary. On this map, "extreme" shows areas with more than five missed intervals.

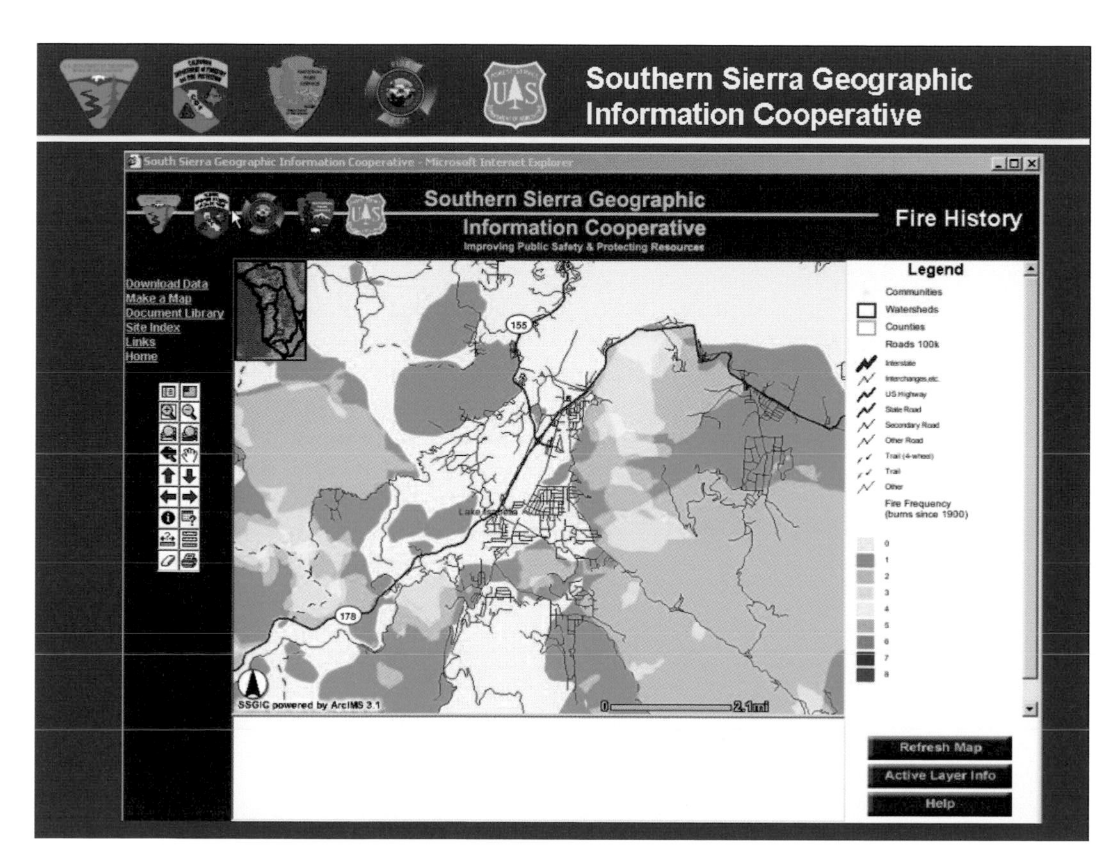

Color coding shows the frequency of fires since 1900 around Lake Isabella in the southern Sierra Nevada range of California. The display helps fire officials plan prescribed burns and other fire management strategies to promote a healthy environment and safe community.

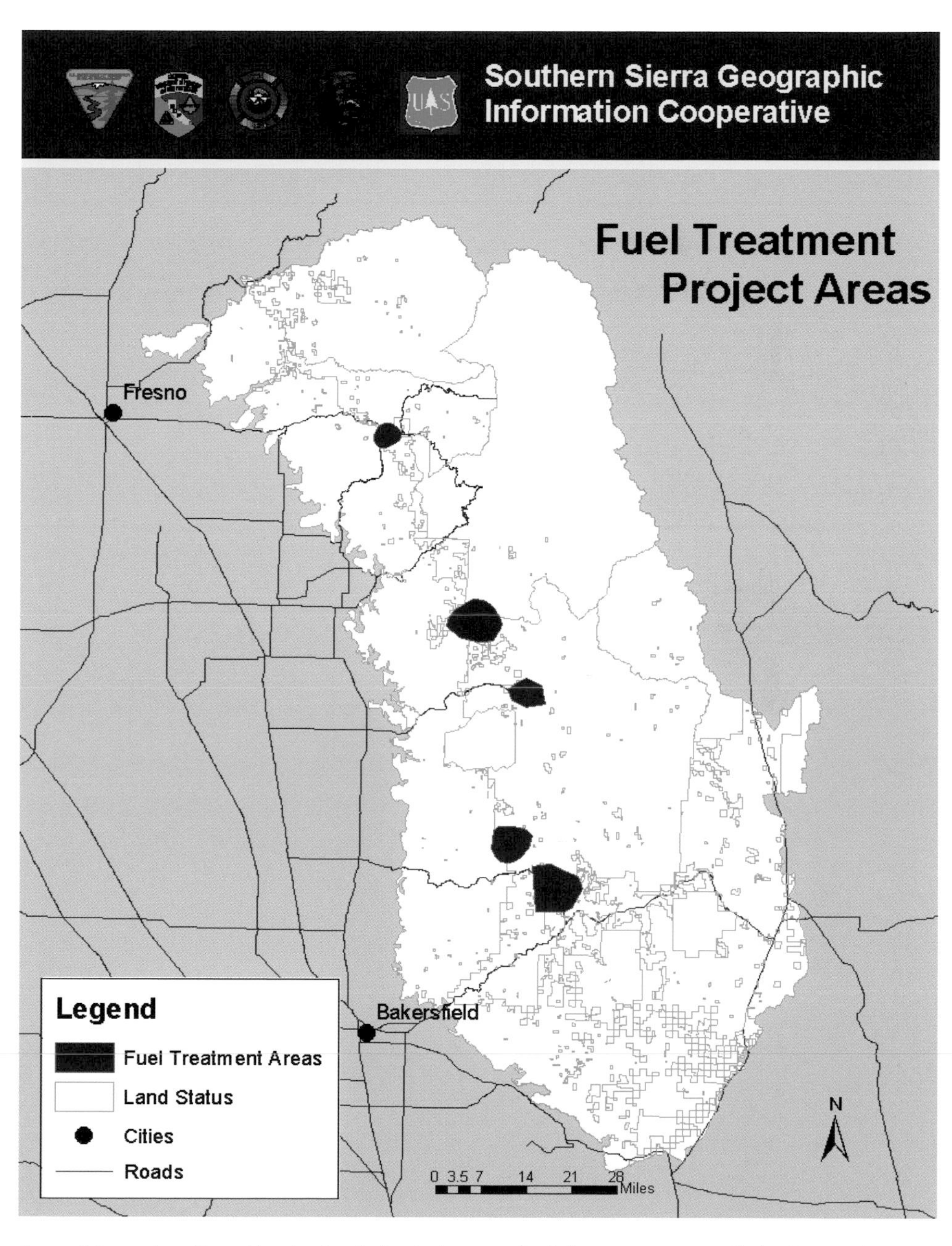

Areas of the southern Sierra Nevada identified as high priority for fuel treatment projects like brush removal, prescribed burns, and other fire management strategies.

Space-Based Burn Severity Mapping in the National Parks

This map captures all of Mesa Verde National Park, as well as the Bircher and Pony fires, within a single subset of a single Landsat satellite scene. It also shows a grayscale burn severity view of the area.

From the saw-grass prairies of Everglades National Park in Florida to the expansive black spruce forest tracts of Denali National Park and Preserve in Alaska, fires affect many different types of vegetation. Fires also burn in many different ways. Fires intensely scorch and radically change some areas while leaving others untouched. Since 2000, the National Park Service has joined several other federal agencies to produce burn severity data sets and map the severity of all large wildland fires in our national parklands with satellite imagery. These GIS-ready maps are valuable because they describe the degree of environmental change caused by fire within a burned area. Burn severity maps help fire managers understand the effects of fire on the forest structure, fuels, and successional patterns of vegetation in the national parks. Combining data collected on the ground and from aerial reconnaissance and satellite imagery, these maps also identify unburned areas within a fire perimeter and measure the likely impact of a fire on grasses, shrubs, and trees in burned areas. In addition, burn severity maps are useful to help guide rehabilitation measures by identifying burned areas vulnerable to erosion and invasion by nonnative grasses and to assess the condition of vegetation and fuels remaining after a fire. Future efforts will focus on interpreting burn severity data and using the information in fire management practices nationwide.

This view of a 1999 fire in Yukon-Charley Rivers National Preserve in Alaska's central region shows fire's heterogeneous effects and a wide assortment of burn severity levels.

GIS combines a map showing burn severity levels of a fire at Yukon-Charley Rivers National Preserve with three photographs that illustrate the effects of the different levels.

Integrated Use of a GIS and the Woodland Home Forest Fire Hazard Rating System

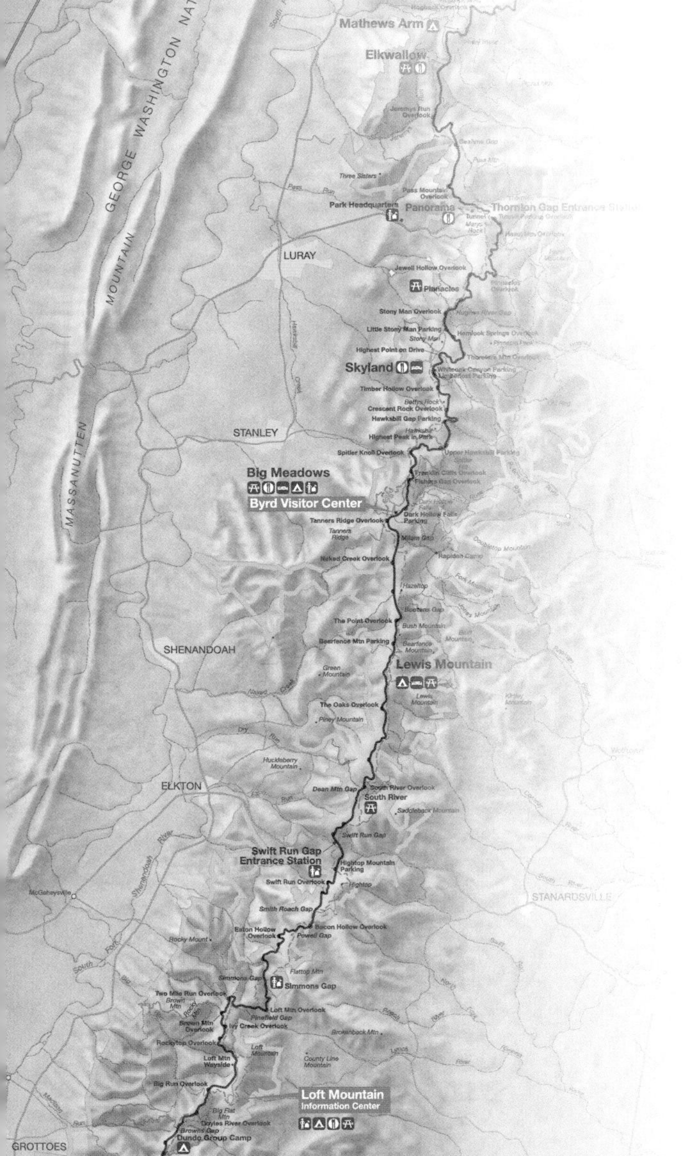

GIS played a crucial role in reducing the fire threat to homeowners living in woodland areas near Shenandoah National Park in Virginia and Acadia National Park in Maine.

The National Park Service uses a program that relies on GIS software to assess the threat of wildfire to homeowners who live at the edge of the two parks. It also works as an educational tool to bring homeowners and fire response managers together with a common goal: reducing the fire danger.

At Shenandoah, fire managers developed GIS themes that included homeowner contact information, fire hazard ratings, water sources, evacuation routes, resource list, neighborhood maps, and fire department information. They have identified neighborhoods at risk of losing homes during a wildfire. This information is now available in the park's GIS system to improve fire education and fire-fighting efforts, specifically in rural areas where homes and nature collide. At Acadia, fire managers are using a similar GIS approach to illustrate potential fire dangers, sending their findings to individual property owners, along with photographs documenting problem areas and suggestions about ways to reduce the fire risk, such as clearing brush from around homes, removing firewood from beneath wooden decks, and removing vegetation around subdivision signs.

Collaboration among the park service, local fire departments, and homeowners will lead to future campaigns to remove fuels that would endanger homes during a fire, enhancing safety for everyone. "Although each approach is unique to their respective programs, there is one common goal," said Dan Hurlbert, a GIS specialist for the National Park Service. "Working as partners with homeowners and community response agencies, we can effectively reduce loss caused by wildland fires."

WOODLAND HOME FOREST FIRE HAZARD RATING

Agency unit code: NPS-ACAD Subdiv_id: 31 Sub-Division Name: Fire Department jurisdiction:
Location: Tax map #: Date established:

Assesment Date
Observer
Lots # Acres:
Homes built:
Homes under construction:

FUEL HAZARD RATING
Choose the predominant type.

Light, low-hazard fuels
(1) (Short grasses, weeds, few shrubs, or mature hardwoods with no understory)

Medium-hazard fuels:
na (2) Woodland:
(3) (Mixed upland forest with fairly open understory with leaf litter and small shrubs)

na (2) Small, flashy fuels:
(3) (Abandoned fields; brush, large shrubs, small trees, cedars, tall grasses)

High-hazard fuels:
(4) Woodland
(Mixed upland forest with heavy large brush, evergreens, downed trees ,limbs and ladder fuels)
(5) Large, flashy fuels
(Evergreen timber stands)

SLOPE HAZARD RATING
Choose the predominant type.
(1) Mid slopes, 0-5%
(2) Moderate slopes, 6-15%
(3) Steep slopes, 16-25%
(4) Extreme slopes, 26% or greater

STRUCTURE HAZARD RATING
Choose the predominant combination of design characteristics.
(Ratings occurring between those shown shall be assigned where they represent areas of mixed structures)
(1) Noncombustible roof and Noncombustible siding materials
(3) Noncombustible roof and combustible siding materials
(7) Combustible roof and noncombustible siding materials
(10) Combustible roof and combustible siding materials

SAFETY ZONE RATING
Choose the predominant range that best represents the precentage of homes that do not have at least 30 feet of defensible space (safety zone) between home and fuel.
(3) 30% of homes
(6) 31-60% of homes
(10) 61-100%

ADDITIONAL FACTOR RATING: ACCESS
Other factors may be permitted in addition to those listed.
MEANS OF ACCESS FOR EMERGENCY VEHICLES
na (3) There is only one means of access into the subdivision
na (3) Road width(s) do not allow 20way emergency vehicle traffic (road surfaces less than 16' wide)
na (2) Road grade(s) are more than 15 percent
na (3) Dead-end roads do not have adequate turn-arounds such as a 100' wide cul-de-sac
na (3) Existing bridge size(s) limit some access of emergency equipmei

ADDITIONAL INFORMATION OR SPECIAL FEATURES: (continue on back)

ADDITIONAL FACTOR RATING: OTHER
na (2) Most roads and streets are not marked with names or numbers on clearly visible signs
na (2) Subdivision entrance is not marded
na (2) Individual home locations or addresses are not marked
na (2) Power lines are not buried
na (2) Area does not have municipal water sources
na (2) Area does not have static water source
na (3) (4) (5) Area is more threatened due to special situations such as high density of homes and distance from fire department

Easy 1...2...3...4...5 Difficult
Plow line:
Hand line:
How easily can area be plowed or raked for fireline, based on steepness or rockiness?:

HAZARD RATING SUMMARY
Fuel Hazard Rating:
x Slope Hazard Rating:
= Fuel x Slope Hazard Rating:
Structure Hazard Rating:
Safety Zone Rating:
Emergency Access Rating:
+ Other Factor Rating:
Fire Hazard Rating:

Risk Level:

Browse to different assesments of a site

National park fire managers use rating systems like this to determine the fire threat to homeowners living at the edge of national parks in the so-called wildland/urban interface.

Vision Fire Burn Chronology

Firefighters trudge through smoke during the 1995 Vision Fire.

The National Park Service turned to GIS technology to show the path and severity of a wildfire that burned more than twelve thousand acres and destroyed forty-five homes in and around Point Reyes National Seashore north of San Francisco.

The park service measured soil types, geology, and vegetation afterward and combined that information with existing GIS data to monitor how the 1995 Vision Fire affected park resources, including forests and grasslands. The park combined data showing the perimeter of the fire with a map of park vegetation prepared before the fire. Putting these two layers of information on a single GIS map clearly identified a mosaic of burned and unburned patches left after the fire. The GIS maps illustrated that most of the fire burned at lower intensities, affecting mostly evergreen shrubs, Douglas firs, and redwoods. The fire burned at higher intensities through Bishop pine forests. Using this information, fire managers developed strategies to rehabilitate burned areas and to understand the patterns of how fires burn and spread in the national seashore area. The park service created a poster from this information to post at trailheads and include in trail guides to help the public understand how nature recovers from fire.

Firefighters ignite a prescribed burn in 1999 as part of the park's fire management.

Point Reyes National Seashore
California

N

Vision Fire Burn Chronology - October 4-6, 1995

October 4, 2:00 a.m.

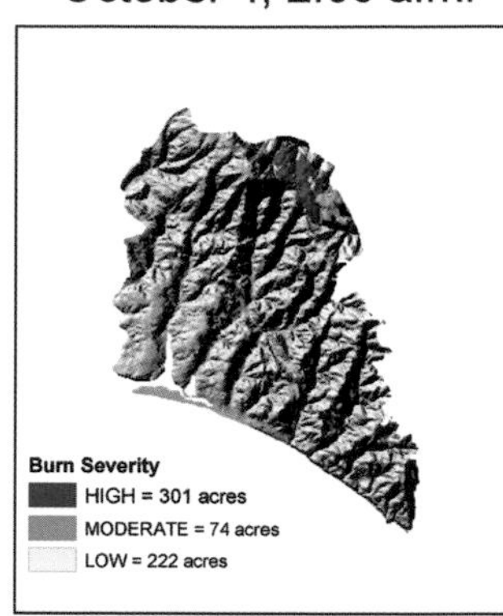

October 4, 3:45 a.m.

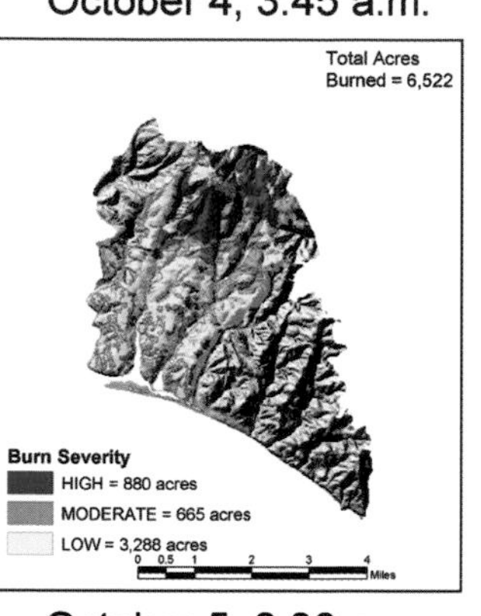

Plant Communities Affected By Vision Fire

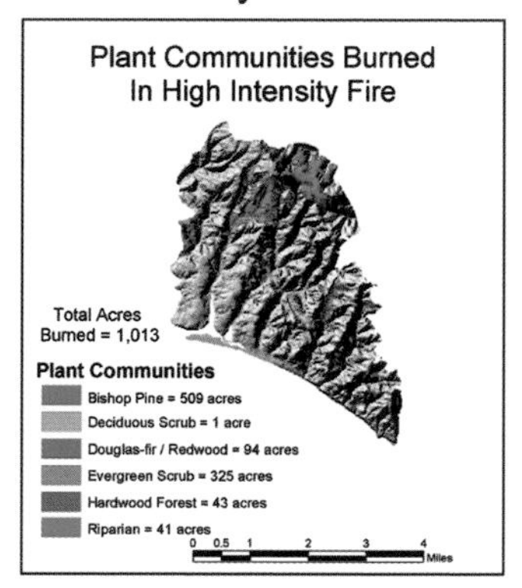

October 4, 5:30 p.m.

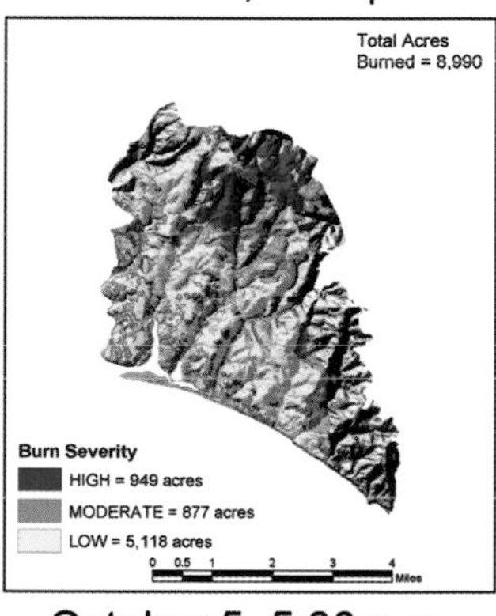

October 5, 8:30 a.m.

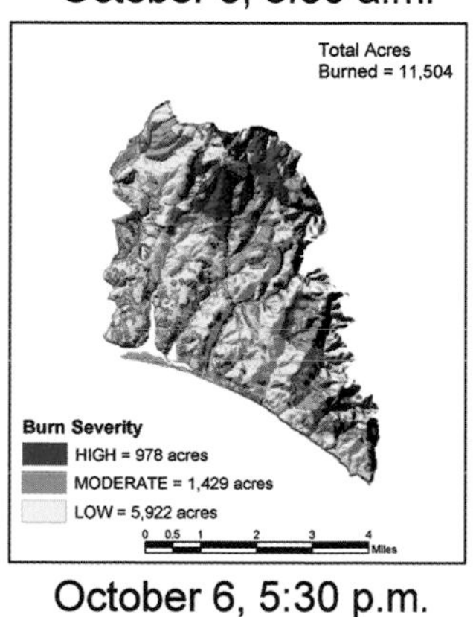

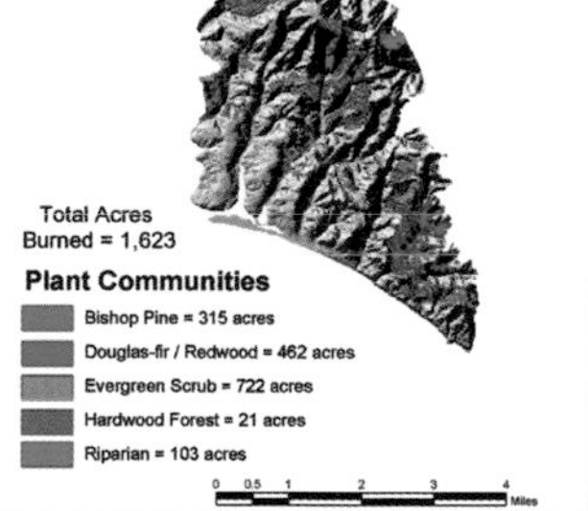

October 5, 5:30 p.m.

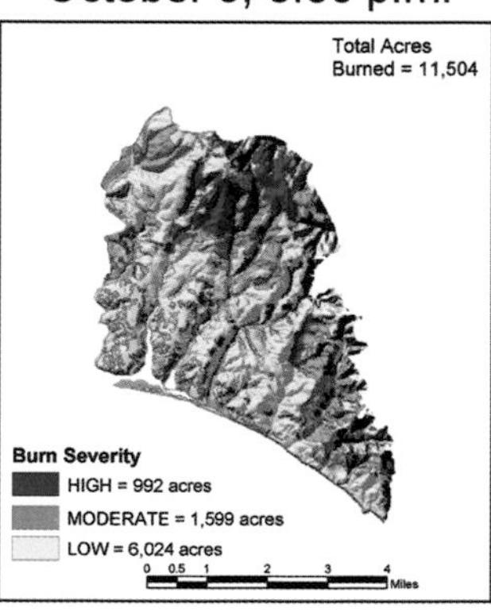

October 6, 5:30 p.m.

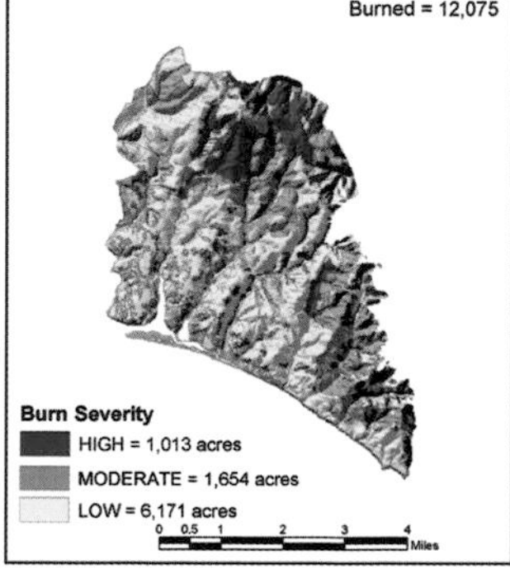

Plant Communities Burned
In Low Intensity Fire

Total Acres
Burned = 6,170

Plant Communities
Bishop Pine = 397 acres
Douglas-fir / Redwood = 622 acres
Evergreen Scrub = 4848 acres
Hardwood Forest = 93 acres
Other = 12 acres
Riparian = 198 acres

Burn intensity and types and total acres of plants burned are shown in this poster chronicling the 1995 Vision Fire.

Field GIS Provides Timely and Accurate Structural Damage Assessment during Wildfire

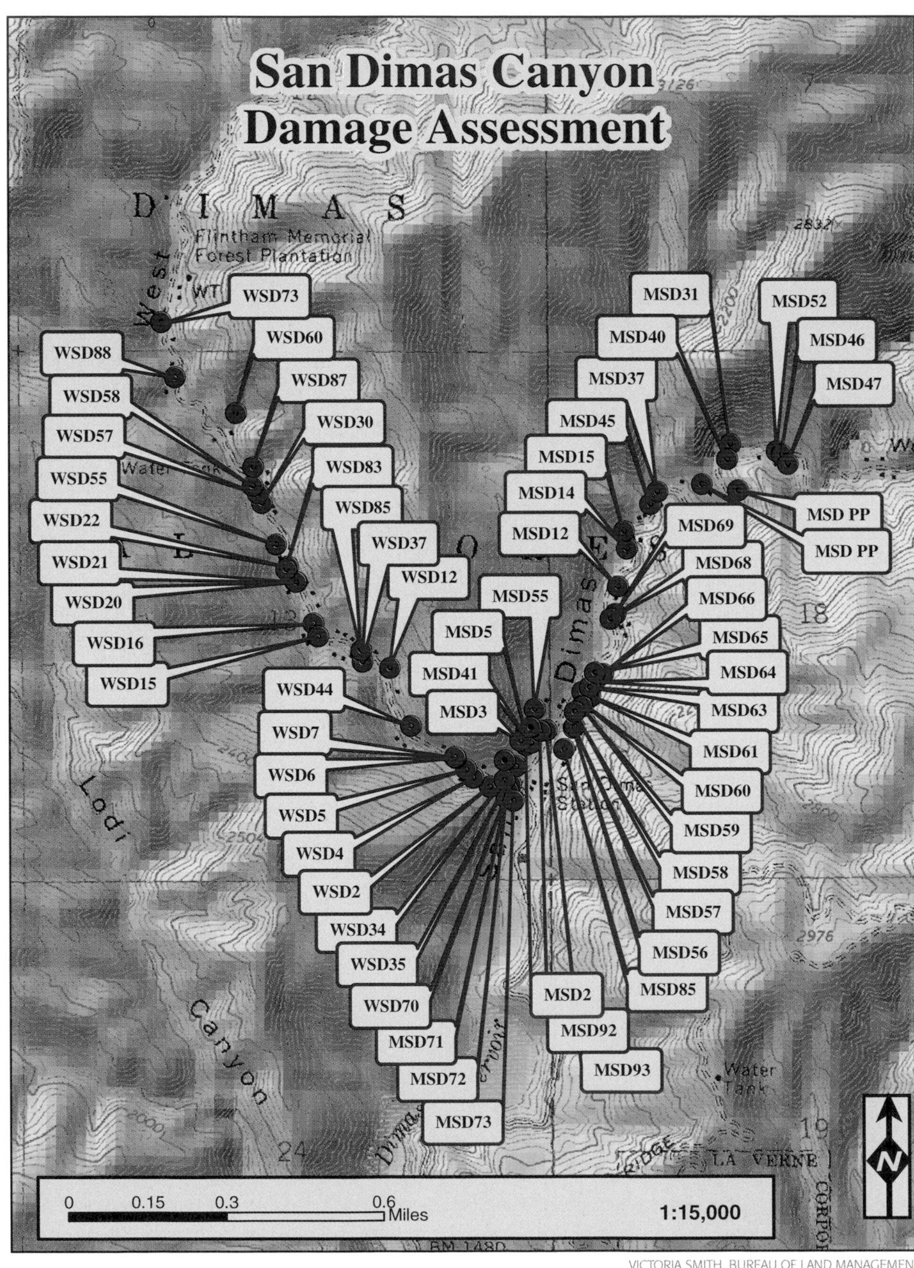

VICTORIA SMITH, BUREAU OF LAND MANAGEMENT

The locations of damaged or destroyed structures in the Williams Fire are identified in West San Dimas Canyon (WSD) and main San Dimas Canyon (MSD). Related numbers represent lot numbers.

In 2001, the Viejas wildfire in California became the proving ground for ArcPad® software, a GIS mapping program that displays fire perimeters in real time. A year later, National Park Service fire managers pushed the limits of ArcPad during the weeklong Williams Fire, which burned nearly thirty-nine thousand acres of the Angeles National Forest near Los Angeles. The blaze became the nation's top fire priority as extreme weather conditions forced the evacuation of thousands of residents and destroyed or damaged more than seventy homes and seasonal cabins. To learn the extent of damage, fire managers supported by GIS specialists sent a team of field observers into the fiery chaos with GIS mapping software, handheld computers, GPS receivers, and digital video and still cameras, along with standard fire-fighting tools and protective gear. The team used ArcPad to navigate through the fire areas and to collect data and display the location of every damaged building and car. The team returned to a nearby ranger station and provided a polished report to fire managers within twenty-four hours, eliminating the time-consuming step of traveling to a more distant GIS lab to translate the data into useful GIS maps. "This would not have been possible without GIS technology," said Tom Patterson, Bureau of Land Management assistant fire management officer in Southern California. Just three years earlier without ArcPad, said Patterson, formerly a fire management officer at Joshua Tree National Park, it had taken fire managers nearly three days to collect and display the data on the Willow Fire, which destroyed sixty structures in the San Bernardino National Forest.

NEWMAN'S I.R. MAPPING

Smoke billows from the 2002 Williams Fire in this aerial photograph that shows burned hillsides and neighborhoods threatened by fire near Los Angeles.

Sparks flash from a flare gun, or "very pistol," as a firefighter sets a backfire in an attempt to consume fuel in the path of the advancing Williams Fire.

An Erickson "Air-Crane" helitanker drops fire retardant during the Williams Fire.

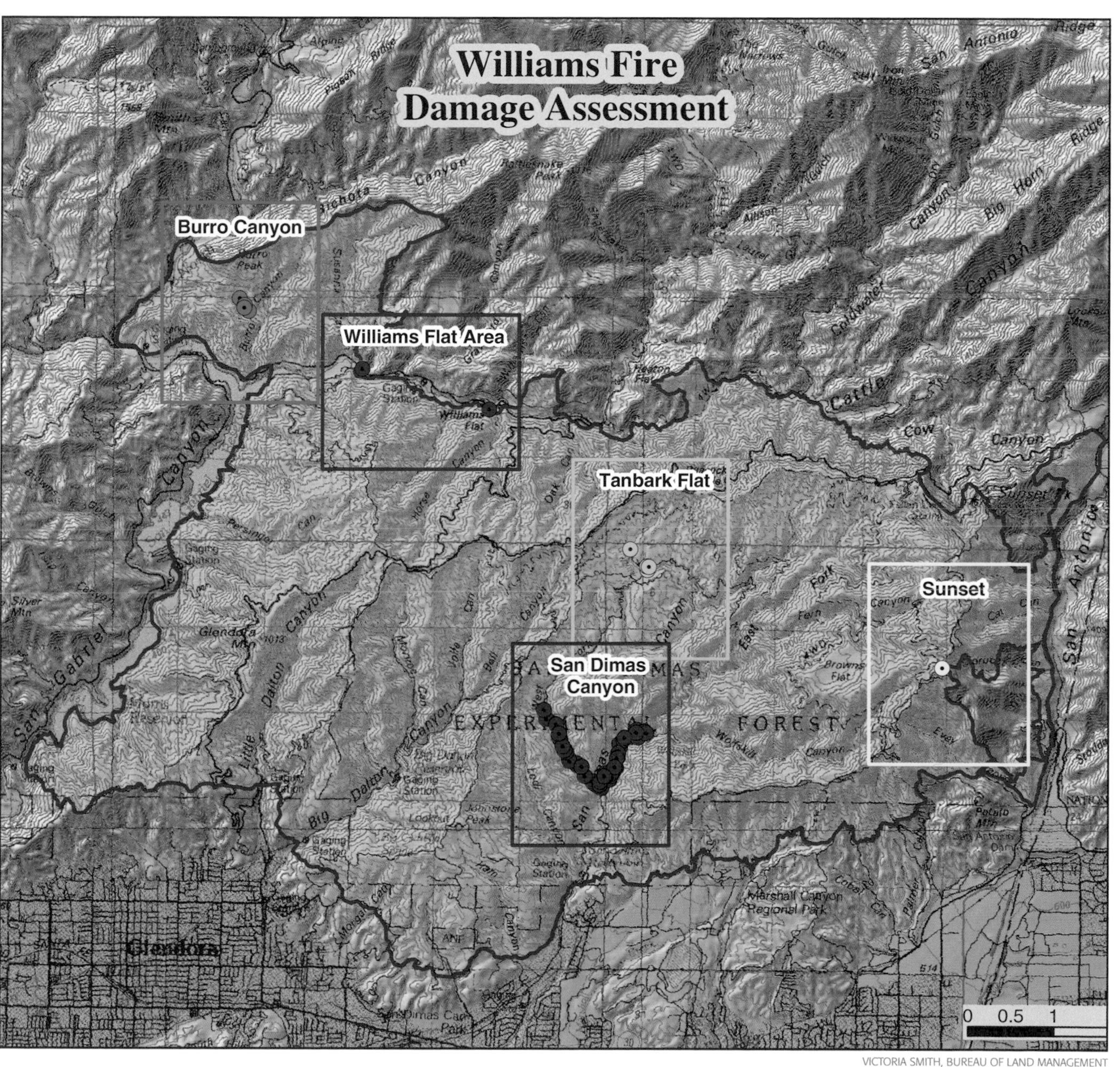

VICTORIA SMITH, BUREAU OF LAND MANAGEMENT

Flames damaged or destroyed homes and other property in San Dimas Canyon and four other areas, as displayed on this GIS map of the Williams Fire.

www.nps.gov/gis

Find and download GIS data about national parks, learn how the National Park Service uses GIS, and more.

maps.nps.gov

Explore America's national parks, "from a national view down through high-resolution views of individual parks," at the National Park Service Interactive Map Center.

www.geodata.gov

The Geospatial One-Stop portal aims to improve government and public "access to geospatial information and data."

www.fgdc.gov

The interagency Federal Geographic Data Committee leads the effort to develop the National Spatial Data Infrastructure (NSDI).

nationalmap.gov

The National Map, an interactive online service from the U.S. Geological Survey, makes geospatial data available to public- and private-sector decision makers.

plasma.nationalgeographic.com/mapmachine

National Geographic's online atlas, MapMachine, lets you "find nearly any place on Earth, and view it by population, climate, and much more" and "browse antique maps, find country facts, or plan your next outdoor adventure."

www.esri.com

ESRI is the world leader in the geographic information system (GIS) software industry.

www.esri.com/esripress

ESRI Press publishes award-winning books about the science, application, and technology of GIS. You can purchase these books directly from ESRI. They are also available at better bookstores worldwide.

Contributors

Chapter 1 *Outdoor Fun: Visitors and Recreation*

Jim Gale, Introduction

Marianna Young, Scenic Quality Analysis in the St. Croix River Valley

Marie Sauter, Hiking Trails of Great Falls, Maryland: Mapping for Visitor Safety

Riley Hoggard and Jeffrey Reed, Personal Watercraft Environmental Assessment: Resource Mapping and Analysis

Tim Smith and Larry Murphy, Use of GIS for Visitor Information at Dry Tortugas National Park

Chapter 2 *Into the Wild: Park Wilderness*

David Graber, Introduction

Joni Piercy, Backcountry Management Planning in Alaska's National Parks

Joni Piercy, GIS Application Development in the Alaska Region

Patrick Flaherty, Wilderness Expansion 2002

Peter Lindstrom, Backcountry Inventory

Chapter 3 *Alive and Well: Park Health and Quality*

John D. Ray, Introduction

Craig Dalby and Mark Stoermer, Puget Sound Basin Air Pollutant Visualization

Jeff Bennett, Katmai Water Resources Management Plan

John D. Ray, Paul Sutton, and Andrew Bingham, Mapping Air Quality Values for the NPS Inventory and Monitoring Program

Mark Lehman, Colonial National Historical Park Elevation

Viktoria Magnis, The U.S. Air Force and National Park Service Western Pacific Regional Sourcebook

Chapter 4 *Without a Trace: Human Impacts*

Janet Cohen, Introduction

Allison Graves, Mapping Subsistence Agriculture in the National Park of American Samoa

Bob Strobe, Kantishna Hills Mining Court Cases, Denali National Park and Preserve

Joni Piercy, Subsistence Hunting in Aniakchak National Monument and Preserve

Chapter 5 *High-Tech Digs: Archaeology*

Steve Baumann, Introduction

Cathy Schwemm and Mike Kaberline, Weights of Evidence Analysis for Cultural Resource Site Prediction and Risk Assessment

Dave Frederick, Clay Tobacco Pipe Stems in New Towne, Jamestown Island

Matthew Wilson, Managing Cultural Resources with GIS: Museum Collections

Steve Klingler, Archaeology and GIS in Western Arctic National Parklands

Chapter 6 *Then and Now: Historic Preservation*

John J. Knoerl, Introduction

Andrew Waggener, GIS in Ruins and Historic Structures Preservation at Salinas Pueblo Missions National Monument

Ann M. Wolley Vawser, Using Historic Maps to Anticipate and Manage Archaeological Resources in Fort Smith National Historic Site, Arkansas

Carl Drexler, Using GIS to Identify Possible Artillery Positions and Manage Artifact Data at Wilson's Creek National Battlefield

Cheryl Sams, Bill Slocumb, and Hugh A. Devine, Linking Cultural Resource Databases through GIS

John J. Knoerl, Mapping the Revolutionary War and War of 1812

Chapter 7 *Into the Future: Park Plans*

Suzy Stutzman, Introduction

Dave Hammond, Land Parcel Prioritizing at a New National Park Service Unit

Greg Jarvis and Julie Seidel, GIS Visualization, Alternatives Development, and Impact Analysis

Jennifer McCollom, Park Planning with Science and GIS

Justin Berthiaume, GIS in the General Management Plan

Kerri Mich, Mapping Proposed Park Expansions

Matthew Safford, Alternative Management Zones for Monocacy National Battlefield

Tammy Stidham, Potomac Heritage National Scenic Trail

Chapter 8 *Safe and Sound: Roads and Trails*

Thomas Fake, Introduction

Kerri Mich, Utility Mapping in the Intermountain Region

Melanie Beck, Using the Power of a GIS Trail Inventory for Trail Planning

Richard Menicke, GIS Database for Rehabilitation of Going-to-the-Sun Road

Riley Hoggard, Habitat Analysis and Road Relocation following Hurricane Opal

Chapter 9 *Leading the Way: Park Maps*

Leslie Armstrong, Introduction

Bill Eichenlaub, Data Access

Dave Frederick, Natural Heritage Occurrences, Colonial National Historical Park

Dave Schirokauer, Enhanced Wetlands Mapping at Point Reyes National Seashore

Jess Grunblatt, Flexible Image Classification with ArcGIS

John Pinamont, Map Products

Thomas Fake, Facility Management in a Historic Settlement

Tim Connors, Tom Clark, Bruce Heise, Bob Higgins, Anne Poole, Stephanie O'Meara, Trista Thornberry, Victor deWolfe, and John Graham, Geology . . . It's Not Just for Scenery Anymore

Chapter 10 *Creature Comforts: Wildlife Management*

Cay Ogden, Introduction

Brian Barns, Determining Foraging and Roosting Areas for Mastiff Bats (*Eumops* spp.) Using Radio Telemetry

Darrell Echols and Donna J. Shaver, Satellite Tracking of Endangered Kemp's Ridley Sea Turtles

Seth P. D. Riley and Denise Kamradt, Effects of Urbanization and Habitat Fragmentation on Bobcats and Coyotes in Southern California

Larry Katahira, Will Seitz, Eldridge Naboa, and George Balazs, Hawksbill Turtles of Hawaii

Ben Dorsey and Tiffany Potter, Lynx in Yellowstone

Neal Darby, A Pilot Geographical Information Systems Assessment of Rocky Mountain Bighorn Sheep Habitat in and around Great Basin National Park, Nevada

Patrick Flaherty, California Condor Viewshed Analysis

Chapter 11 *The Big Picture: Ecosystems*

Andrew Valdez, Introduction

Ann Hebig, 2002 Exotic Plant Management: Chemical Control

Frank Partridge, South Florida Composite Topographic Mapping

Greg Daniels, Gregory Nowacki, Page Spencer, Michael Fleming, and Torre Jorgenson, Unified Ecoregions of Alaska

Mark Adams and John Portnoy, Restoration of Historically Restricted Estuaries, Cape Cod National Seashore, Massachusetts

Phoebe Vanselow and Lewis Sharman, Alaska Coastal Resources Inventory and Mapping Program

Chapter 12 *Setting GIS on Fire: Managing Fire*

Brian Sorbel and Brad Cella, Introduction

Brian Sorbel, GIS: Tool for Fire Managers

Ann Birkholz, Using GIS to Support Collaborative, Landscape-Level Fuels

Brian Sorbel, Space-Based Burn Severity Mapping in the National Parks

Dan Hurlbert, Integrated Use of a GIS and the Woodland Home Forest Fire Hazard Rating System

Erin Noonan, Kate Levendosky, and Dave Schirokauer, Vision Fire Burn Chronology

Tom Patterson, Field GIS Provides Timely and Accurate Structural Damage Assessment during Wildfire

Image Credits

The editors gratefully acknowledge the contributions of the individuals and organizations listed below. Other photos and images in this book appear courtesy of the National Park Service.

ON THE COVER Joshua Tree National Park
Photodisc.

ON THE TITLE PAGE Winter in Zion National Park
Copyright © 2001 Photo 24. Brand X Pictures.

CHAPTER INTRODUCTION PHOTOS

Chapter 1: Riding stock in Grand Canyon National Park
Copyright © 2001 Photo 24. Brand X Pictures.

Chapter 2: Rain forest in Olympic National Park
Copyright © 2001 Photo 24. Brand X Pictures.

Chapter 3: Great Smoky Mountains National Park
Copyright © 2001 Photo 24. Brand X Pictures.

Chapter 4: Barn in Grand Teton National Park
Copyright © 2001 Photo 24. Brand X Pictures.

Chapter 5: South rim of Canyon de Chelly National Monument
Copyright © 2001 Photo 24. Brand X Pictures.

Chapter 6: Buildings in Tumacácori National Historical Park
Copyright © 2001 Photo 24. Brand X Pictures.

Chapter 7: Crystal Forest in Petrified Forest National Park
Copyright © 2001 Photo 24. Brand X Pictures.

Chapter 8: Overhead view of Sequoia National Park
Copyright © 2001 Photo 24. Brand X Pictures.

Chapter 9: Map of Yellowstone National Park
Library of Congress Geography and Map Division, Washington, D.C., 1886, scale: 1:125,000, 90 × 66 cm.

Chapter 10: Wheeler Peak in Great Basin National Park
Copyright © 2001 Photo 24. Brand X Pictures.

Chapter 11: Grand Canyon of the Yellowstone
Copyright © 2001 Photo 24. Brand X Pictures.

Chapter 12: Aerial view of smoke from Williams Fire, Angeles National Forest near Los Angeles, September 25, 2002
Newman's I.R. Mapping.

PAGE

3 Camping at Mt. Jefferson. Copyright © 2000 C. Borland/PhotoLink. Photodisc.

4 Bald eagle. Copyright © 2000 Alan and Sandy Carey. Photodisc.

4 Man and woman canoeing. Copyright © 2000 Photodisc.

4 Pylon and power lines. Copyright © 2000 Skip Nall. Photodisc.

8 Dolphin. Copyright © 2000 Lawrence M. Sawyer. Photodisc.

8 Man on a jet ski. Photodisc.

10 Couple snorkeling. Copyright © 2000 S. Pearce/PhotoLink. Photodisc.

13 Couple resting from a hike. Copyright © 2000 C. Borland/PhotoLink. Photodisc.

16 Mountains in Denali National Park and Preserve. Tom Wiley, NPS Photo.

16 Caribou grazing in Denali National Park and Preserve. Tom Wiley, NPS Photo.

20 Jackson Lake in Grand Teton National Park. Copyright © 2001 Photo 24. Brand X Pictures.

23 Traffic jam caused by visitors viewing wildlife, Gardner Canyon, Yellowstone National Park. NPS Photo.

24 Mount Rainier. Copyright © 2001 Photo 24. Brand X Pictures.

26 Valley of Ten Thousand Smokes, Katmai National Park and Preserve. David Duran, NPS Photo.

26 Boy with fish. Copyright © 2000 PhotoLink. Photodisc.

27 Grizzly bear. Copyright © 2000 PhotoLink. Photodisc.

32 Jet fighter. Copyright © 2002 Stocktrek Corporation. Brand X Pictures.

35 Covered wagons at Eagle Rock, Nebraska. Copyright © 2000 C. Borland/PhotoLink. Photodisc.

35 Pioneer farmstead. Copyright © 2000 PhotoLink. Photodisc.

38 Denali National Park and Preserve. Tom Wiley, NPS Photo.

38 Panning for gold. Copyright © 2000 C. Borland/PhotoLink. Photodisc.

40 Leopard seal resting on ice. Copyright © 2000 PhotoLink. Photodisc.

44 Male elk. Copyright © 2000 PhotoLink. Photodisc.

46 Pipe stems recovered during archaeological excavations in historic Jamestown. Colonial Williamsburg Foundation.

46 Park GIS Specialist Dave Frederick using a handheld GPS unit to gather data about federally protected species in Colonial National Historical Park. Charles D. Rafkind, NPS Photo.

47 Bricks outline an archaeological site in historic Jamestown. Charles D. Rafkind, NPS Photo.

47 Dr. Audrey Horning of the Colonial Williamsburg Foundation at historic Jamestown excavation. Colonial Williamsburg Foundation.

48 Chaco Anasazi bowl in Chaco Culture National Historical Park collection. Puerco black-on-white, from Pueblo Alto site, 1030–1200 A.D. NPS Photo.

48 Chaco Anasazi corn-grinding tools in Chaco Culture National Historical Park collection. Metate (underneath): Trough slab for grinding corn into flour. Mano (resting on metate): Two-handed grinding tool. NPS Photo.

62 Hand holding a global positioning system unit. Copyright © 2001 Joaquin Palting. Photodisc.

65 Mountain bike and cyclist on overlook. Photodisc.

70 Cougar. Copyright © 2000 Stocktrek Corporation. Photodisc.

76 Gettysburg Battlefield. Copyright © 2000 C. Borland/PhotoLink. Photodisc.

81 Two hikers at Arches National Park. Copyright © 2000 Karl Weatherly. Photodisc.

81 Road in Arizona. Copyright © 2000 F. Schussler/PhotoLink. Photodisc.

84 Wildflowers on a Grand Canyon trail. Copyright © 2000 Ken Samuelsen. Photodisc.

91 Couple with map. Copyright © 2000 Photodisc.

92 Humpback whales breaching. Copyright © 2000 PhotoLink. Photodisc.

92 Glacier Bay, Alaska. Copyright © 2000 J. Luke/PhotoLink. Photodisc.

93 Alaskan black bear. Copyright © 2000 PhotoLink. Photodisc.

94 Powhatan Creek, Colonial National Historical Park (photo at upper left). Charles D. Rafkind, NPS Photo.

PAGE

94 Rare skipper butterfly. Anne Chazal, Virginia Department of Conservation and Recreation, Division of Natural Heritage.

94 Virginia state zoologist Christopher Hobson surveying park resources. Virginia Department of Conservation and Recreation, Division of Natural Heritage.

96 Yellow bush lupine at Great Beach, Point Reyes National Seashore. Sue Van Der Wall, NPS Photo.

97 Moonrise at Drakes Beach, Point Reyes National Seashore. Sue Van Der Wall, NPS Photo.

97 Lighthouse at Point Reyes National Seashore. David Duran, NPS Photo.

98 Polar bear. Copyright © 2000 Geostock. Photodisc.

100 Alaskan glacier. Photodisc.

102 Hawaiian cliffs and coast. Photodisc.

104 The Castle, Capitol Reef National Park. Copyright © 2001 Photo 24. Brand X Pictures.

105 Head frame at Duchess Mine, Capitol Reef National Park. John Burghardt, NPS Photo.

107 Bighorn ram. Copyright © 2000 PhotoLink. Photodisc.

107 Buffalo at Yellowstone National Park. Copyright © 2000 S. Alden/PhotoLink. Photodisc.

112 Coyote. Copyright © 2000 PhotoLink. Photodisc.

114 Excavating hawksbill nest on Kamehame Beach. NPS–HAVO Turtle Project.

115 Volunteer prepares to mark nest site. NPS–HAVO Turtle Project.

115 Volunteers identify postnesting hawksbill. Larry Katahira, NPS Photo.

115 Hawksbill prospects for nesting site at Apua Point, Hawaii Volcanoes National Park. NPS–HAVO Turtle Project.

115 Hatchlings scramble toward ocean at Punaluu. Will Seitz, NPS Photo.

115 Volunteer with turtle. Larry Katahira, NPS Photo.

115 Hatchling heads to ocean at Keahou Beach, Hawaii Volcanoes National Park. Will Seitz, NPS Photo.

116 Yellowstone National Park. Photodisc.

123 Mount Rainier and alpine meadow. Copyright © 2000 PhotoLink. Photodisc.

123 Nisqually Glacier, Mount Rainier National Park. Copyright © 2001 Photo 24. Brand X Pictures.

124 Devils Tower National Monument. Copyright © 2001 Photo 24. Brand X Pictures.

126 Aerial view of Florida wetlands. Copyright © 2000 C. McIntyre/PhotoLink. Photodisc.

127 Alligator. Copyright © 2000 C. McIntyre/PhotoLink. Photodisc.

128 Mist at Kenai Fjords National Park. Copyright © 2000 PhotoLink. Photodisc.

135 Pulaski ax photo courtesy of California Department of Forestry/Riverside County Fire Department, Indio Station 86.

135 Laptop computer. Photodisc.

135 Satellite. Photodisc.

140 Forest after a fire (photo at left). Photodisc.

142 Skyline Drive in Shenandoah National Park. Copyright © 2001 Photo 24. Brand X Pictures.

143 Acadia National Park. David Duran, NPS Photo.

146 San Dimas Canyon Damage Assessment Map. Victoria Smith, Bureau of Land Management.

146 Aerial view of Williams Fire, Angeles National Forest near Los Angeles. Newman's I.R. Mapping.

147 Williams Fire Damage Assessment Map. Victoria Smith, Bureau of Land Management.

Index